Customer-Centered Telecommunications Services Marketing

For a listing of recent titles in the *Artech House Telecommunications Library,* turn to the back of this book.

Customer-Centered Telecommunications Services Marketing

Karen G. Strouse

Artech House
Boston • London
www.artechhouse.com

Library of Congress Cataloging-in-Publication Data
Strouse, Karen G.
  Customer-centered telecommunications services marketing / Karen G. Strouse.
  p. cm.—(Artech House telecommunications library)
Includes bibliographical references and index.
ISBN 1-58053-854-1 (alk. paper)
1. Telecommunication—Marketing. 2. Customer relations. 3. Consumer satisfaction.
4. Competition, International. I. Title. II. Series.

HE7631.S758 2004
384'.043'0688-dc22

2004052339

British Library Cataloguing in Publication Data
Strouse, Karen G.
  Customer-centered telecommunications services marketing—(Artech House telecommunications library)
  1. Telecommunication—Marketing 2. Relationship marketing
  I. Title
  384'. 0688

ISBN 1-58053-854-1

Cover design by Gary Ragaglia

© 2004 ARTECH HOUSE, INC.
685 Canton Street
Norwood, MA 02062

All rights reserved. Printed and bound in the United States of America. No part of this book may be reproduced or utilized in any form or by any means, electronic or mechanical, including photocopying, recording, or by any information storage and retrieval system, without permission in writing from the publisher.
  All terms mentioned in this book that are known to be trademarks or service marks have been appropriately capitalized. Artech House cannot attest to the accuracy of this information. Use of a term in this book should not be regarded as affecting the validity of any trademark or service mark.

International Standard Book Number: 1-58053-854-1

10 9 8 7 6 5 4 3 2 1

To Art

Contents

	Preface	xiii
1	**The State of the Telecommunications Industry**	**1**
1.1	The Need for Customer-Driven Telecommunications Marketing	1
1.2	The State of the Telecommunications Marketplace	3
1.3	The Customer and the Distribution Channel	6
1.4	The Customer and the Decision-Maker	8
1.5	Customers and the Regulatory Environment	11
1.6	Customers and Technology	12
	References	13
2	**The Nature of Telecommunications Competition**	**15**
2.1	Market Structure	15
2.2	Marketing Telecommunications Services as a Commodity	20
2.3	Differentiating Telecommunications Services	22
2.4	Emerging Marketing Opportunities and Strategies of Industry Leaders	24

		References	26
3		**The Marketing Plan for a Customer-Centered Telecommunications Enterprise**	**27**
	3.1	The Marketing Planning Process	27
	3.2	Directing the Marketing Plan to the Customer	29
	3.3	Differentiation Strategies	32
		References	38
4		**Market Research**	**39**
	4.1	Data Mining and Customer Profiling	39
	4.2	Primary and Secondary Market Research	43
	4.3	Yield Management	47
		Reference	50
5		**Customer Demand**	**51**
	5.1	Capital-Intensive Industry Demand	51
	5.2	Technology Industry Demand	55
	5.3	Price Elasticity and the Demand Curve	56
	5.4	Forecasting Demand	58
		References	62
6		**Competitive Intelligence**	**63**
	6.1	Developing Competitive Intelligence	63
	6.2	Finding Strategic Competitors Before the Customer Does	68
	6.3	Perceptual Mapping	71
	6.4	Competitive Parity	74
		References	75
7		**Channel Management**	**77**
	7.1	Traditional Distribution Channels	77
	7.2	Evolving Routes to Market	78

7.3	Cross-Channel Strategies	81
7.4	Partner Relationship Management Systems	83
7.5	E-Business Channel Management	84
7.6	Strategic Alliances	86
7.7	Scoping the Channel as a Customer-Centered Strategy	87
	References	89

8 Stakeholder Management — 91

8.1	The Importance of Stakeholders	91
8.2	Customers as Stakeholders Throughout Changing Regulations	96
8.3	Issues Management	97
8.4	Crisis Management	100
	References	102

9 Market Segmentation — 105

9.1	Markets and Submarkets	105
9.2	Statistical Segmentation Using Cluster Analysis	110
9.3	Market Coverage Strategies	111
9.4	Customer Lifetime Value Management	113
9.5	Making the Most of Undesirable Markets	116
	References	116

10 Pricing — 117

10.1	Avoiding Commodity Pricing	117
10.2	Pricing Strategies	119
10.3	Promotional Pricing	122
10.4	Bundling	123
10.5	Price Discrimination	125
10.6	Pricing for Competitive Parity	127
	References	129

11	**Customer Loyalty, Retention, and Churn**	**131**
11.1	Predicting and Limiting Churn	131
11.2	Strategies for Maintaining Loyalty	134
11.3	Limiting the Costs of Churn	137
11.4	Benefiting from Churn	140
	References	142
12	**The Customer Relationship Management Process**	**143**
12.1	Customer Relationship Management Systems	143
12.2	Using Billing Systems to Enhance Customer Relationships	147
12.3	Customer Self-Service	150
12.4	One-to-One Marketing	153
	References	155
13	**Branding**	**157**
13.1	The Importance of the Brand	157
13.2	Brand Equity	160
13.3	Product Positioning	162
13.4	Branding Commodities	164
13.5	Protecting the Brand	166
	References	169
14	**Marketing-Based Innovation**	**171**
14.1	Distributed Innovation	171
14.2	Marketing Disruptive Technologies	173
14.3	Maintaining an Innovative Organization	177
14.4	Case Study: NTT DoCoMo's i-Mode	180
	References	182

15	**Customer-Centered Technology Marketing**	**183**
15.1	The Uniqueness of Technology Services Marketing	183
15.2	Revisiting the Diffusion of Innovation Curve	188
15.3	Technology Marketing Tools	191
	References	194
	Acronyms	**195**
	About the Author	**197**
	Index	**199**

Preface

Customer-Centered Telecommunications Services Marketing

It is always fun to follow and then write about the telecommunications services industry, but it's also a challenge to write about an industry when you know that it will change dramatically as soon as you have committed the analysis to the publication process. While I was writing a book about strategy in 2000, a proposed merger between WorldCom and Sprint fell apart. I'd been extolling the strategic value of the combination, but regulators clearly didn't share my views. During the writing of this book, Cingular bought AT&T Wireless. Once burned, twice shy: I no longer assume that any events will unfold as we expect them to unfold, even after agreements are signed. Maybe that merger will complete, and maybe it will succeed. Whatever happens, it is likely some other momentous consolidation that took place during the publication process is conspicuously absent from these pages.

Daily events redirect our expectations in the same way a one-degree miscalculation might change the trajectory of a tossed ball. The evolutionary events surprise us, too, and that same one-degree miscalculation over the course of two decades might change as the view through a telescope. Anyone who worked for then-AT&T as I did in 1982 remembers the moment they heard about the consent decree that was to break up the company on the first Orwellian day of 1984. We talked about competition that day, and "fully separated subsidiaries," and that AT&T would threaten IBM's lock on the computer and equipment business as IBM was leering at the long-distance market. In fact, some of AT&T's and IBM's most embarrassing failures were in each other's industries within a few years of our speculation. But we missed the biggest news about both of them. Neither IBM nor AT&T is the colossus it was in those days.

Long distance and equipment were the only pieces of the industry we ever discussed, other than the paper directory business. Most of us ignored or were simply ignorant of today's most vibrant industry segments. I never imagined that the first video we'd see on a consumer phone would be on a wireless one, though I might have guessed that it would happen in Japan. My voice messages and faxes arrive in my e-mail inbox, and I buy my electronics on-line before I drive around the corner to pick them up or check their delivery status with overnight services. Tomorrow's ordinary activities are almost unimaginable.

Yet, at the same time the industry seems to move so quickly, in other ways it travels like molasses. Twenty years after deregulation got started around the world, competition is fierce in some sectors and still nearly absent in others. Real competition in local service is just beginning to peer up out of the ground in most countries, though it seems as though this time it is on its way here. But it has never been more evident that customers are really in charge now.

Customers decide which brand is the leader, no matter how much energy and money service providers invest in trying to persuade them. Customers seek out the best service quality, looking to customer survey research, not technical studies. In other words, customers want to use the service provider that other customers, not engineers, have decided has the best quality. Customers choose who will lead the market, what service providers will sell, and why they buy.

It's an irony that an industry that established itself as all about reaching out and talking has learned that it needs to listen. Like marketing, pricing, and distribution, even listening has changed. It's not enough to ask customers if they are happy; happy customers churn at an alarming rate. Asking customers what they want in the future will mire the service portfolio hopelessly in the past. Customers do not seek out the service provider that knows what they want; they seek out the one that knows what they will want. Service providers need to be flexible and innovative, getting to market before someone else gets the chance, with a service that works at a surprisingly low price. It is a challenge, but someone always meets it. The goal of this book is to help its reader be the service provider who does.

I would like to take this opportunity to thank all of my colleagues, clients, and mentors who have helped me think about this exciting marketplace, and the Artech House staff and reviewers who provided so much support and valuable improvements during the writing, editing, and production processes.

1

The State of the Telecommunications Industry

1.1 The Need for Customer-Driven Telecommunications Marketing

A telecommunications executive once said [1], "Operations looks from equipment, out. Marketing looks from the customer, in." In a nutshell, knowing the difference between operations and marketing is the key to competitive advantage. Those who know the difference and conduct business accordingly are still in the minority.

The landscape of telecommunications service providers has changed dramatically during the past quarter century, but it is still a hodgepodge of real competition and residual monopolies or near-monopolies. Service lines that do not bear the vestiges of regulation—wireless, Internet access, and long-distance services—are intensely competitive. Mature service lines that were once offered only as monopolies have become somewhat competitive while only a few consumer monopolies—fixed-line local access, cable—have remained relatively impervious to the vortex of competition swirling around them.

The long-awaited convergence of services, such as cable telephony or broadband entertainment services to the home, has not yet arrived, but its imminence has changed how the future competitors plan and operate. There are already hints that markets will be more transparent and customers will have more choices. In the United States, customers can transfer their landline numbers to a wireless phone, then hop from service provider to service provider at will without the cost of switching. The owner of a service provider in one country might well be the diversified owner of a provider halfway around the world. Deutsche Telekom, once the government-owned and operated monopoly in

Germany, purchased the largest all-GSM wireless service provider in the United States and named it T-Mobile.

Some of the venerable incumbent service providers have shrunk and become focused, or disappeared into mergers, while yesterday's startups have turned into today's full-service behemoths. Rival service providers have matured in the face of intense competition. The recent slump experienced in the global telecommunications services industry coincided more directly with the overall economic environment than any previous time in history.

The growing capabilities of transmission media and a surge in network investment created a glut of capacity just in time for the economic slump. Monopolies that manage the network planning for an entire industry rarely have to cope with excess capacity and stranded investment, but competitive service providers are not protected from their own inaccurate forecasts. In other words, the telecommunications industry is showing signs that it will operate in the future just like the competitors in any other industry. All of these apparently disparate trends signify that the rules of competitive markets finally govern most of the telecommunications services industry.

Before competition ruled the market, telecommunications service providers directed their focus on the needs of regulators and less directly, the needs of investors. Indeed, the path to higher returns and happier investors was often through the regulatory process. Those customers who were the most exploited, euphemistically described as profitable—businesses, consumers in urban centers, and subscribers to enhanced services—often found their requests eclipsed by the most vocal nonspenders who bothered to complain to local regulators. A primary focus on paying customers was not only unproductive; it was almost considered tawdry.

To be fair, few businesses in any industry valued a customer focus until the last decades of the twentieth century. In the earliest demonstration of the modern retail environment, Henry Ford's famous quote, "Any color, as long as it's black" demonstrated the manufacturer's reluctance to alter its marketing approach to meet the diverse needs of customers. Technology advances that enable considerable customization of products and services, coupled with the blurring of differences between telecommunications and entertainment services, all have contributed to the ability of service providers to meet the diverse needs of customers.

Customers have become more educated and more demanding as well. In a global economy increasingly characterized by service industries shaped by technology, customers are impatient for services customized to their individual needs and expect them to be offered at lower prices with each new rollout. Customer-centered marketing, once a differentiation tactic for telecommunications providers, has become a survival strategy. After all, most of what is known as marketing happens inside the customer's mind.

1.2 The State of the Telecommunications Marketplace

In the United States, though most of the market is still moving towards full competition, certain submarkets have effectively arrived. Examples include wireless services, long distance, and Internet access. Markets that are still on their way are broadband data services and business access. Markets for which there is still no effective competition include facilities-based narrowband local access and in-home entertainment services.

Nonetheless, the notion that an abundance of local service providers would emerge from AT&T's dismantlement has not materialized. Of the eight major local service providers in place at divestiture in 1984, there are four remaining. Verizon and SBC each comprise three of the original regional companies, and they might now be described as super-regional, in the case of SBC, and national, in the case of Verizon, which now includes the GTE network. US West was absorbed into Qwest, and BellSouth remains unmerged, though it has established significant marketing partnerships such as the Cingular wireless brand.

According to the U.S. Federal Communications Commission (FCC), by the end of 2002, competitive local exchange carriers supplied 25 million local lines in the United States to the incumbent carriers' 163 million, about 13% of all switched access lines. Competitive local exchange carrier switched access lines increased in 2002 by 26%.

Most of these facilities were resold or leased under a program that offers unbundled network elements for resale. While the mandated regulatory program has been successful at increasing the presence of additional local carriers in U.S. markets, true competition requires genuine facilities-based network alternatives. On the other hand, resale creates an apparently competitive marketplace in which marketing is among the most important differentiators. Furthermore, resale offers an entry strategy for companies whose success could eventually fund the construction of local facilities. Only about 26% of competitive access lines are facilities owned by the competitive service provider.

The U.S. long-distance market shrank slightly between 2001 and 2002, reflecting lower prices per minute and an abandonment of the conventional network to other, uncounted transmission alternatives. Though AT&T still dominated the market in 2001, its share dropped from 90% in 1984 to less than 38%, while MCI boasted a 24% share, Sprint had 9%, and the remaining 24% of the market was divided among more than a thousand competitors. Service providers providing contractual residential long-distance services found their customer base shrinking as consumers gravitated to dial-around services, cable and Internet-based services, prepaid services, and bundled wireless plans.

In the 27 largest U.S. markets, wireless services grew at 5.3% between 2002 and 2003, a drop from its 7.6% rate between 2001 and 2002, according to

J.D. Power and Associates. Churn rose in the same period from 24% to 26%, and was expected to rise again when local number portability went into effect in November 2003.

In September 2002, the U.S. Commerce Department's Technology Administration found that the U.S. penetration of consumer broadband services, at 10.4%, lagged behind South Korea (51.7%), Hong Kong (26%), Canada (19.7%), Taiwan (18.2%), and Sweden (13.4%) [2]. Signifying a marketing problem rather than one of broadband availability, by the end of 2002, one estimate found that 90% of U.S. subscribers had broadband access at the same time, though only 31% could choose between digital subscriber line (DSL) and cable alternatives. The same study showed that 48.7% of businesses had some kind of broadband service.

The European Union issued a Framework Directive in 2002 in an effort to standardize the liberalization process among its member countries. While implementation was supposed to occur by mid-2003, most member countries had not yet complied when the date passed. The guidelines intended to identify the level of competitiveness in each industry segment, remove regulatory restrictions on those markets adequately governed by market forces, and oversee those markets that remained noncompetitive. While the logistics of implementation were complex, the simple existence of such a directive demonstrates the drive to deregulate markets in an orderly yet definitive manner. By the end of 2001, mobile penetration in Europe as a region was 75% [3].

According to Insight Research [4], cultural factors and the fact that Europe lags the United States in technology adoption will hasten the growth of Wi-Fi hotspots for wireless local access. Europe's less mature office PC market will create a Wi-Fi penetration through newer, Wi-Fi–equipped PCs rather than upgrades to old systems. Europe's more prevalent use of wireless services, dense populations, older buildings that are difficult to wire for conventional networks, mass transit systems, and a café culture will all serve to foster Wi-Fi implementation and adoption.

As late as 2003, Germany's Deutsche Telekom had held competitors to 4.4% of fixed-line connections, though competitors had acquired a higher amount of market share in large cities [5]. Italy's incumbent local provider still had 71% of the market in 2002, rising due to competitive failures and lackluster deregulatory initiatives [6], with less than 1% of the total fixed lines in the country resold through Italy's unbundling program. While Italy's telecommunications users stayed well ahead of most countries in mobile services, Internet service lagged other European countries. By 2002, Internet penetration was at 20 million customers.

The United Kingdom implemented the European Union's Framework Directive in July 2003, and though some services remain regulated for the near future, the United Kingdom is among the more competitive local service markets. Residential broadband service, for example, is cheaper in the United Kingdom than in the United States, France, and Germany, second in affordability only to Sweden in a recent benchmarking study [7].

France Télécom claimed 80% market share in 2002 in the local-calling market and retained about two-thirds of the national and international calling volume [8]. French subscribers tend to lag the rest of Europe in technology adoption, and by 2002, penetration in the broadband market was about 5%. Though other European countries have higher mobile penetration than France's 2003 level of 65.3%, three service providers account for the mobile market. Spain and the United Kingdom, for example, each boasts a mobile penetration rate of 80%, and Italy's mobile penetration rate is 91%. About nine-tenths of Italy's mobile users are prepaid.

The Scandinavian markets of Sweden, Denmark, and Finland were among the first to comply with the European Union's deregulatory framework. A 2002 review of broadband in Sweden concluded that penetration rates of about 15% were the highest of any Western nation. Furthermore, by 2001, Sweden's mobile penetration rate was already at 46.4% [9]. Sweden deregulated its cable television service in 1992. By mid-1999, 70 companies offered cable television services in Sweden.

Russia's mobile market grew 124% in 2002, and its overall market by 40%, owing to deregulatory measures and the effect of the emerging market economy. At a fixed-line penetration of 14% in 2003, the market was far from saturated at the January 2004 effective date of the new regulatory environment [10].

Asian markets are highly competitive in wireless, and well ahead of the rest of the world in deployment of broadband and mobile technology. Japan has the lowest DSL prices in the world [11] and led the world in DSL subscribers at the end of 2003. According to DSL research firm Point Topic, Japan boasted 9.2 million subscribers, followed by the United States with 8.2 million and China with 7.8 million DSL subscribers. Japan's enviably developed mobile market offered 3G technology when most of the developed economies were rolling out GPRS. Eighty percent of Japan's cellular subscribers held Internet subscriptions by the end of 2002. China's telecommunications market is divided nearly in half between fixed-line and mobile users, though the mobile calling segment grows faster than fixed-line calling [12]. South Korea's broadband penetration of 73% was the highest in the world in early 2004, with 60% connecting through DSL and 36% using cable broadband.

The mobile market in Australia is among the most competitive in the world. Federal and state governments have invested about 10 million dollars to build the first 3G wireless network in the southern hemisphere. Located in Adelaide, the network is used as a testbed for new services and applications [13]. Hutchison Telecom launched its commercial 3G network, simply called 3, soon afterwards.

In Latin America, wireless penetration exceeds fixed penetration from a two-to-one ratio to a six-to-one ratio [14]. After industry consolidation, four wireless carriers serve 85% of the Latin American market. While DSL is the common broadband application in the region, incumbent providers provide 85% of the market.

Service providers operating in a variety of telecommunications sectors or across geographical markets face varying states of competition, growth rates, and levels of market penetration. The challenge is that a customer-focused perspective often means a multitude of approaches to meet the needs of many very small segments of customers, not a unilateral market. Segmenting into business and consumer markets to define customer targets might be suitable for a mission statement, but not for a marketing plan. Thus, a unilateral customer focus is a losing approach in a competitive environment. The service provider that meets most of the needs of all customers will find itself with a very small constituency.

1.3 The Customer and the Distribution Channel

Before service providers can fine-tune their offerings to very individualized needs, strategic marketing decisions do apply to broad customer segments. From the moment competition was introduced, service providers who were owners of large capital assets have tangled with their own important reseller customer as a result of their competition for the service provider's other important retail customer. The network asset owners were nearly always the incumbent local service providers, either telecommunications or cable companies, who viewed resellers of their services as unwelcome but mandatory-to-serve cream-skimmers that narrowed markets, reduced profits, and complicated commerce. The resellers traditionally viewed these suppliers as obstructionist and inefficient. To some degree, they were both right. Nevertheless, participants in this unpleasant relationship recognize that the situation will last only until regulators step out of the management of the industry and the competitors can narrow each other's markets, squeeze margins, and complicate commerce on their own terms.

Channel conflict, or disintermediation, occurs when suppliers form relationships with their customers' own customers. For example, a clothing manufacturer with its own stores might compete against department stores for customers. It becomes a conflict if the stores are within the shopping range of the same customers, or if the manufacturer sells directly to consumers at a discounted price, or if the manufacturer provides services that the department store cannot match. The Internet is undoubtedly responsible for widespread disintermediation in the retail sector, with on-line customers able to purchase directly from manufacturers, bypassing the distributor. While this disintermediation is common across most industries, incumbent telecommunications service providers, on account of their vast and underused asset bases, risk alienating large and profitable customer segments when they operate aggressively in both wholesale and retail channels.

Most customers are largely unaware of an interchannel squabble until it affects their own service. Nonetheless, informed customers are well aware of the choices they have among service providers. Customers and service providers sustain opposite desires in the number of channels available to them. Customers prefer diversity, and would most appreciate a variety of service providers, more than the market warrants, because that would produce a multitude of service choices and pressures to keep prices low. A market with too many participants would weed out underperformers in the long term, but customers would have the flexibility to move from one service provider to another, or to maintain accounts with many service providers as loss-leader pricing occurs in one service line or another.

Distribution channels for all but a few telecommunications service providers are all but predestined. Incumbent carriers owning and operating billions of dollars in plant assets are wholesalers, either de facto or de jure. Building a network is very expensive. New entrants to the market can limit their geographical coverage to a comparatively infinitesimal size or decide to resell the facilities owned by a large incumbent. The adaptation of older technologies such as cable or satellite or the buildout of new technologies like fixed wireless will introduce facilities-based competition to the market. Until viable, full-coverage alternatives are in place, conflicts are unavoidable. But the kind of customer focus needed to succeed in a vibrant, competitive market will restructure the industry, whether by choice or by legally mandated structural separation. In a less regulated, more open market, wholesalers might choose to offer their retail services in different markets than do their own customers. When operating in different physical territories is not viable in many local markets, serving customer segments not targeted by one's own resellers resolves the channel conflict without precluding all retail markets.

Customer-centered service providers excel. To do so, service providers need to meet or exceed the already formidable network reliability metrics, and much more. They need to offer simple billing formulas, stated on clear, timely notices. Customer service and support, already a baseline requirement for all telecommunications service providers, offers a telecommunications service provider the opportunity to excel by providing capabilities beyond those within the reach of competitors.

Service providers in the retail channel can offer features, bundles, or innovative pricing algorithms that appeal to customers with unique needs. Wholesale service providers need to compete for customers in a different way. The customer of wholesale services is probably a distributor of access or transmission capacity or a business with its own telecommunications management organization. Most often, service bundles, pricing beyond discounts for very large volumes, or unusual features are not competitive differentiators in the wholesale market. Instead, wholesale customers will seek access to network management tools, flexible reliability levels and pricing to match, and technical customer care with skills well beyond those in the administrative and billing call center.

A brave industry observer could argue that the current crop of service providers, or even one service provider, excels at marketing in the retail channel. On the other hand, few would credit the incumbents who serve both the retail market and the wholesale channel with excellence in innovative wholesale channel marketing. With few exceptions, the service providers in the wholesale channel operate, if not under protest, then at least without enthusiasm. The lack of competition and alternatives for customers discourages the retailer/wholesaler from enhancing the wholesale network with customer-accessible network management features. Furthermore, the complex regulatory formulas and procedures limit the wholesaler's ability to control pricing, and incumbents often complain that the pricing in place does not cover their costs. Lastly, incumbents make the argument that unprofitable or at least obligatory service provision discourages new network investment that would be resold at a loss.

Telecommunications service providers presently do a better job of meeting the distribution needs of the channels that were dealt to them than the channels they are competitively positioned to serve. A customer-centered telecommunications service provider will revisit its distribution channel profile and seek excellence through its customer's eyes.

1.4 The Customer and the Decision-Maker

Buyer markets divide rather neatly between those in which the end user of the service is the buyer and those where buying responsibilities lie with an expert with accountability for the purchase decision. End users who make purchase decisions are ordinarily consumers or behave like consumers. Decision-makers who represent the interests of the user or a community of users are ordinarily business buyers. Normally, the larger the purchase, the more senior the decision-maker will be. Large enterprises often keep a permanent technical staff that advises the decision-makers of the technical merits of the telecommunications services alternatives. But the final decision often lies with the chief executive officer, the chief information officer, or the chief financial officer. Mid-sized and small businesses that cannot justify full-time technical expertise sometimes hire consultants to gain the expertise they need for purchase decisions only on a temporary basis. Government purchases sometimes involve inviolate decision-making criteria not in the control of any employee or committee involved in the purchase. Consumer users most often are the decision-makers for their own telecommunications services, though innovative marketing techniques that have been borrowed from other industries could transform that paradigm.

There are many differences between consumers and business buyers, and most arise from the fact that the business buyer is representing a user or class of users, and paying with somebody else's money. Even a business owner makes decisions more like a business buyer at work, and more like a consumer at home. Similarly, business purchases of items and services of a value under some predefined threshold behave like the purchasing behavior of consumers. Even a cost-conscious business cannot be bothered to invest in a feasibility study of directory assistance providers. The threshold level is different between enterprises and households, and between one enterprise and another. Even so, most business purchasing is above the threshold requiring analysis and most daily consumer purchasing falls below it. This distinction translates into a set of characteristics.

The business buyer is knowledgeable about telecommunications services, features, and benefits. The business buyer expects the service provider to provide the information needed to produce an analysis evaluating the purchase on its financial merits. The rewards must be tangible and quantifiable. Most often, and depending on the size of the purchase, the analysis compares the cost of the alternatives and the benefits solely in terms of their monetary value. If intangible rewards are very large, the cost/benefit analysis will attempt to quantify them, but in most cases, the analysis simply mentions and then ignores them. Consumers, on the other hand, often overlook the differences in operational costs of two competing purchases, or they might fail to quantify the benefits they gain from each alternative as compared to their incremental costs. Consumers might enter into a contractual wireless agreement on the basis of an attractive promotional handset with little consideration for the usage costs throughout the life of the contract. A business buyer is more likely to calculate the estimated monthly charges of competing plans, the availability and cost of technical support, and the value of the features compared to products with fewer features and lower cost.

Well-managed enterprises measure the quality of business purchases as part of the buyer's performance. Thus, business buyers look for cost effectiveness rather than simply low price in their purchase decisions. Furthermore, business buyers seek to reduce their personal risk. The reputation of the vendor is far more important than the lowest price, especially in large telecommunications investments that are mission-critical to the enterprise. Consumers are more unpredictable than business buyers in this regard. They are more apt to suffer the inconvenience of churning their long-distance providers to attain a constant level of promotional awards. On the other hand, the same consumer might also buy new technology at a premium just to be among the first to have it.

Business buyers are buyers first, and initiate the evaluation of services only when they believe that they are ready to purchase and can find a service that will meet their needs at a justifiable price. Business buyers are aware of time as a cost factor. The astute business buyer might hasten to install a new cost-reducing technology though its price might drop in a year, or delay its installation until its track record and subsequent price reduction produces more cost justification and less risk. Many consumers shop less for the purpose of buying and more for the entertainment value it provides. A consumer might delay buying even if a product meets their decision criteria at a cost-justified price. Consumers might also buy products that well exceed the price of alternatives, to meet personal needs that are unrelated to the capabilities of the purchase and their predefined needs.

Selling to business often involves direct sales, multiple contacts, and negotiations with several or many representatives of the organization. Large procurements often involve committees and long lead times. Large purchases are complex, require technical assistance, and complicated contractual arrangements. Decision-makers representing multiple business functions complicate even decisions that produce large savings for the enterprise. The solution that creates profit in one organization but effort without return in another will be embraced by the profitable group and resisted by the other. In contrast, consumers make decisions alone and sometimes impulsively. A consumer's purchase of a house might take less time or attention than a business might spend on the purchase of a cleaning service, let alone an office building. Typically, while there is overlap between telecommunications services offered to consumers and those targeted to business buyers, the list of business-only services includes those that are the most complex and the most expensive. Time will tell whether they will also be the most profitable in the long term, though historically business services were indeed the drivers of profit under social policy-led pricing.

Both the business buyer and the telecommunications service provider would like the business relationship to endure. The buyer wants a long relationship, because change is costly. The buyer also knows that each of the service providers competing for the contract is eager to create a long-term revenue stream. Thus, the competing service providers are often willing to invest in marketing toward winning the sale. Consumer sales are smaller, simpler, and require less of a relationship with the service provider during the selling process. Consumers do not seek a long-term relationship with the service provider. On the contrary, they prefer to have the flexibility to switch service providers. Similarly, the service provider would rather not make a large investment to gain the sale, especially because consumers are very likely to churn.

While most decision-making structures are in place to ensure that the needs of the enterprise take priority, conflicts between enterprise needs and end-user desires can adjust the decision process. This is especially likely when the decision-maker does not have accountability for the expenditure. Employees who have discretion over their own vendor selection decisions, even within an approved set of choices, sometimes make their selections using both practical criteria and other parameters without benefit to the enterprise. Airlines have benefited when employees manipulate route planning to aggregate the maximum number of personal frequent flier miles, or when employees select a vendor who offers prizes that cannot be shared with the enterprise. Whether telecommunications service providers can benefit from mismatches of corporate and individual desires—and whether they risk long-term corporate relationships if they do so—requires the strategic attention of management.

1.5 Customers and the Regulatory Environment

Generally, telecommunications services customers were unaware of the complicated foundation of regulations that supported the monopolistic provision of local and long-distance services. No doubt consumers would not have been willing to delve into the policies and calculations underlying prices anyway, nor would such a study give them much satisfaction from its logic. The pricing model for monopoly services was widely quoted by engineers as "measured with a micrometer, then hit with an axe." The pricing met social needs, not commercial ones.

It was once common for consumers of regulated services to believe that their underpriced local service was too expensive and their long-distance service offered at a large multiple of its cost was a great bargain. Thus, one advantage of a competitive market with multiple providers—whether regulated or not—is that customers can judge the value of a service against an alternative service. But even after some services underwent deregulation, their complicated pricing structures remained in place longer than the regulatory framework. For example, though long-distance services had been competitive for about 20 years, it took Sprint until 1995 to offer a per-minute, all-day, distance-insensitive price for a long-distance call within the United States. Until then, customers could only estimate roughly the price of the call they were about to dial.

Consumers of monopoly telecommunications services—or in industry vernacular, residential subscribers—made their purchase decisions based on ability to buy or a subjective assessment of a given service's value. As there were not any competitive alternatives to the services offered by the incumbent, guessing was about as scientific as anyone could be. Business customers could at least compare telecommunications services costs from one year to the next, or compare the price of a phone call to the cost of a customer visit, or value the benefit of the call in dollars.

1.6 Customers and Technology

Much research has explored the relationship between customers and technology. A theory called *diffusion of innovation*, developed at Iowa State College in 1957 and refined by Everett Rogers, identified five categories of adopters to define when in the product life cycle individuals are receptive to new technologies. It is revealing to note that the theory did not arise out of the technologies of the day that come to mind: televisions, computers, or appliances, but rather the adoption by farmers of hybrid seed corn. Rogers, and all the researchers that have followed this field of inquiry for the next half-century, learned that individuals with a similar adoption pattern also differ from the other adopters in their information needs, their risk profiles, their expertise, and their loyalty.

Figure 1.1 demonstrates the pattern of adoption for new technologies. The first type of adopter is the innovator. These individuals are curious and adventurous, enthusiastic about technology, and aware about the latest advances. Innovators are willing to take risks and live with the consequences. Most often the innovator has access to larger-than-average financial resources and a willingness to pay a premium to be progressive. The innovator, though representing only a small percentage of buyers, serves the valuable function of introducing the technology into the view of the remaining customers.

Early adopters occupy a more prominent role than innovators in disseminating new technologies. First, they physically outnumber innovators. More importantly, while buyers considering a new technology view innovators as purchasers of technology for its own sake, they pay more respect to the opinions of

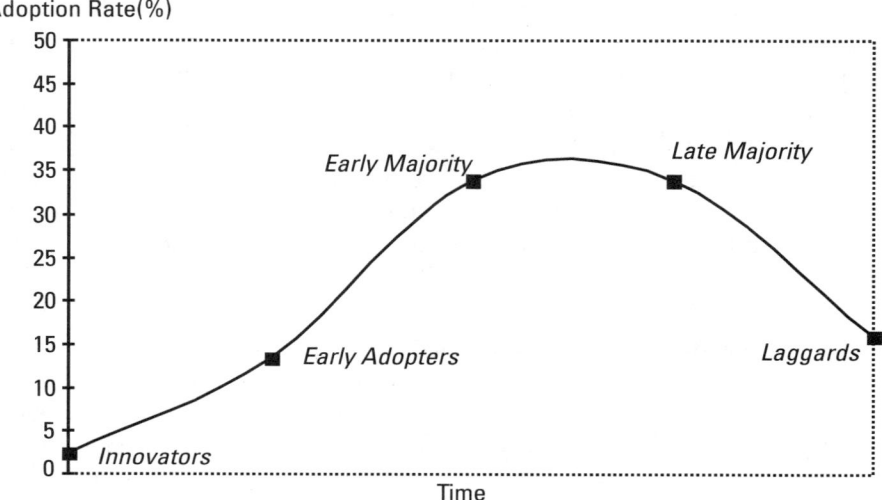

Figure 1.1 Diffusion of innovation.

early-majority adopters. The early adopter often has data from the experiences of the innovators, and a thought-out decision that led to their purchase. The early adopter has the vision to grasp and communicate the service's strategic advantages. Thus, the second wave of buyers will serve as opinion leaders to create growth in the market.

Early-majority and late-majority adopters make up about two-thirds of the market, and require much more data about the benefits of a technology before they will adopt it. While the early-majority individuals are primarily pragmatic, the late majority is cautious. The laggards are simply skeptical about change. Bringing the early- and late-majority adopters into the customer base is crucial, as full penetration of the innovators and early adopters nets only one-sixth of the market. Because the majority adopters are less knowledgeable about technology in general and the new technologies in particular, they require a better infrastructure than do the innovators and early adopters: training, support, and customer care.

Though the categories are useful in evaluating the profiles of buyers, actual individuals do not rest on a given marker, but along the continuum. Furthermore, an individual who is an innovator in one market, such as mobile services, might be in the late majority in another segment, such as broadband.

The concept of a tipping point arose from diffusion of innovation research. The tipping point occurs when adoption (or infection, or the galvanization of a population) reaches a certain critical mass. Marketers should be interested in the tipping point; it transforms a market segment into a revenue epidemic. Telecommunications service providers have dreamed of the revenues from services not yet at the tipping point: on-line movies and entertainment, video conversations, wireless data, and telemetering, but none have yet materialized. But the signs that they are emerging are in data- and camera-enabled mobile phones, broadband deployments, and increasingly intelligent devices and appliances. The service provider that predicts and markets the technology that leads to the tipping point will earn a windfall of profits from innovators and early adopters, and gain a strong market share when the majority follows their leadership. Market research firm comScore announced that more than half of the U.S. population, or 150 million users, used the Internet in a single month during 2003. Wireless penetration in Finland was at 60% by the third quarter of 1999 [15]. By 2002, developed European countries boasted 70% penetration, while the U.S. market lingered at 45% [16].

References

[1] Reeves, Betsy, "Growing Your Data Warehouse," *Wireless Review*, Vol. 15, No. 20, 1998, pp. 24–28.

[2] Piven, Joshua, "Broadband Lagging in US," *Computer Technology Review*, Vol. 22, No.11, 2002, p. 4.

[3] Taaffe, Ouida, "Yankee Predicts Positive Future for European Mobile," *Telecommunications*, Vol. 37, No. 4, 2003.

[4] Insight Research, "Wi-Fi in North America & Europe: Telecommunications' Future 2003-2008," October 2003.

[5] Taaffe, Ouida, "Why It's Called Deutsche Telekom," *Telecommunications International Edition*, Vol. 37, No. 10, 2003, pp. 22–23.

[6] Cazzani, Stefano, "More Revenue, Less Competition," *Telecommunications International Edition*, Vol. 37, No. 10, 2003, pp. 24–26.

[7] McClelland, Stephen, "Paradise Postponed?" *Telecommunications International Edition*, Vol. 37, No. 10, 2003, pp. 30–32.

[8] Wieland, Ken, "The Fall and Rise of France Telecom," *Telecommunications International Edition*, Vol. 37, No. 10, 2003, pp. 16–18.

[9] Malim, George, "A Consumer-Oriented Climate," *Telecommunications International Edition*, Vol. 35, No. 10, 1999.

[10] Pankratova, Oksana, and Jason Smolek, "From Russia with VAS," *Telecommunications International Edition*, Vol. 37, No. 10, 2003, pp. 36–38.

[11] McClelland, Stephen, "Playing Broadband Catch-Up," *Telecommunications International Edition*, Vol. 37, No. 10, 2003, pp. 56–57.

[12] Sun, Lin, "A Changing Landscape," *Telecommunications International Edition*, Vol. 37, No. 10, 2003, pp. 42–45.

[13] McKenna, Ted, "Testing, Testing, 3G Down Under," *Telecommunications Americas Edition*, Vol. 37, No. 1, 2003, p. 8.

[14] O'Keefe, Sue, "Broadband and Wireless Lead the Way Forward," *Telecommunications International Edition*, Vol. 37, No. 10, 2003, pp. 62–64.

[15] Martinek, Marcia, "From West to East?" *Wireless Review*, Vol. 16, No. 18, 1999.

[16] Luna, Lynnette, and Kelly Carroll, "Fourth-Quarter Subscriber Numbers Disappoint Investors," *Wireless Review*, Vol. 19, No. 2, 2002, pp. 10–11.

2

The Nature of Telecommunications Competition

2.1 Market Structure

An assortment of competitive markets faces all but the smallest or most focused service providers. Nonetheless, most telecommunications sectors display some common structural characteristics. The first is that telecommunications services have the characteristics of what is often called "the new economy," the shift that took place during 1975 from a workforce whose time was consumed producing goods and services to a majority of workers who sell, manage, analyze, support, or innovate. Most of the productivity gains that led to the new economy were technology based. Indeed, the brightest examples of new economy enterprises were technology companies themselves. Those enterprises not only produced technology for others, but most of them also built or restructured their own business operations more effectively with technology support.

Operating and improving in the midst of the technology revolution, advanced information technologies have been part of the infrastructure of the telecommunications services industry on both the supply side and the demand side of the market. On the supply side, telecommunications service providers have benefited more than many century-old industries by deploying computer and communications technologies in their service lines and in their back offices. On the demand side, telecommunications services, managed nearly end-to-end by intelligent information systems, are customizable, and that means that customers drive the demand.

The competitive markets model developed by Michael Porter [1], as shown on Figure 2.1, demonstrates the characteristics that are driving

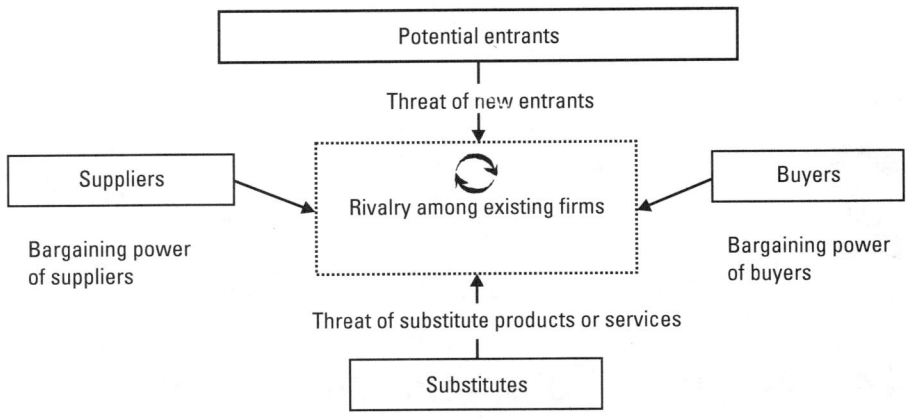

Figure 2.1 Porter competitive markets model.

competition in the telecommunications services industry. The model, applicable to any commercial industry, depicts five forces that define the intensity and nature of competitive behavior. It is imperative for the telecommunications services marketer to view this model from the customer's point of view and respond with services packaged to maintain market share and fend off the efforts of competitors.

The first force is the *intensity of rivalry* among the firms in the marketplace. If a few providers dominate an industry's market share, the industry is regarded as concentrated. Competition in concentrated industries is less intense than if many service providers command leadership in the marketplace, because the few leading enterprises can control the main attributes of the market. It takes a concentration ratio of about 90% to qualify an industry as a monopoly.

The U.S. Census Bureau published its latest concentration ratios for U.S. industry in 1997. One measure of industry concentration—and hence competitive intensity—is the market share held by the top four service providers. At the time, the top four cable networks held 61.1% of the market, the top four fixed-line telecommunications service providers held 47% of their market, and the top four cellular service providers held 51.4%. Though these ratios drop considerably as deregulation takes hold, and indeed these ratios might have dropped in the years since the survey in 1997, these markets are still considered relatively disciplined. In other words, the leading service providers can control the market through pricing, managed technology innovations, service, or other means of differentiating oneself from one's rivals without undermining the basic industry structure.

The leading service providers in concentrated markets have some control over their customers, as long as its largest rivals are willing to cooperate at a pace that falls short of collusion. Thus, when a wireless provider lowers per-minute prices, sooner or later the rest of the leading service providers follow with their own price reductions. Or one of the market leaders offers an innovation that draws customers from the others, such as Nextel's push-to-talk features. The other leaders copy the innovation and if they are successful, market share returns to what it was before the introduction of the innovation in the first place.

But market leaders within this game cannot simply stop lowering prices or innovating, because they cannot control the pace of innovation initiated by their competitors outside the inner circle or its maverick competing industry leaders. Even in concentrated markets, when one of the leaders chooses to change its price significantly, or offers an important differentiating feature, its rivals are forced to respond, and this heightens the intensity of competition. The less concentrated the market, the less an individual service provider can control, or even persuade, its customers.

A concentration ratio under 40% qualifies an industry as highly competitive. The reseller industry earned a rating of 16.6% in the U.S. Census Bureau analysis, signifying a fiercely competitive market. Customers benefit the most from highly competitive industries, enjoying the narrowest margins and thus the lowest prices, and a rapid introduction of innovation.

Small service providers need to fight intensely for customers in virtually any market. The telecommunications services industry features several characteristics that exacerbate the intensity of its rivalry. High fixed costs create a high breakeven point, and the firms already in the market fight fiercely to improve the utilization of their assets. Low switching costs for customers to change service providers, coupled with commodity or near-commodity services, create a high annual rate of customer churn. The high perishability of services—the inevitability that an unsold minute-of-use is gone forever—fosters intense competition, because both the purchase and the timing of the purchase of services are at stake.

While regulation no longer serves as a *barrier to entry* to the telecommunications services industry in most markets, the high price of infrastructure keeps many potential new entrants out of the facilities-based sectors. The same economies of scale that create intense competition also serve as a barrier to entry keeping out new entrants who cannot make the large investments or who cannot sustain a long startup period without profits. By preventing newcomers from becoming profitable until they boost their market share and asset utilization above a breakeven level, the high entry cost deters some would-be entrepreneurs from even entering the marketplace and protects those competitors already operating.

Furthermore, much of the required investment is in application-specific equipment and construction, such as switches, custom software, and transmission equipment. If the venture fails, the startup cannot readily convert the specialized assets to an alternative use. Service providers already in possession of specialized assets also fight to maintain their market positions, because they, too, cannot sell or convert the assets if the business fails. Few investment barriers exist in resale, leading to a marketplace with many competitors, intense competition, and lower margins. The resale market stays in equilibrium because potential competitors enter when profits rise and resellers leave the industry when they fall.

Another barrier to entry is the patent and to a lesser extent the trademark, and telecommunications service providers are intensely protective of services or features that differentiate their offerings from those of their competitors. After the 1984 breakup of AT&T, a negotiation and legal battle ensued as to which of the entities would be entitled to the ubiquitous blue graphic of a bell, a deeply engrained asset of marketing goodwill. The local access providers won the right to the bell and the Bell name, to great fanfare of their victory. Within less than 20 years, little remained of the Bell name or the Bell logo among the marketing weapons deployed by any of the contemporary local entities. BellSouth retained the symbol in its name, and SBC Communications retained only the first letter for the acronym of its corporate parent.

Suppliers place competitive pressures on the markets. The suppliers for telecommunications services include equipment manufacturers, labor, or, in the case of resellers, facilities-based service providers. The limited number of equipment manufacturers creates supplier power. If the service provider is one of many customers of the supplier, or represents a small portion of revenues, suppliers gain power. Of course, the suppliers in one industry are the buyers in another. The slump experienced by switching equipment manufacturers, partly in reaction to telecommunications service providers reducing their capital expenditures, reduced the supply of equipment manufacturers at worst, and reduced supplier investment in research and development at best.

For resellers, supplier power can also arise in the form of disintermediation, that is, their suppliers sell to their customers. While disintermediation affects most industries, resellers of telecommunications services compete against suppliers whose market share in their own territories is generally a large multiple of their own. From the customer's perspective, disintermediation simply creates more choices of service providers, and the reseller that adds value to the service above the competitive price will win the customer from the supplier.

Convergence in the industry has been expected to make the threat of *substitutes* a formidable problem for telecommunications service providers, but markets sometimes evolve against expectations. Analysts predicted for a decade that U.S. cable market shares would fall when satellite television became viable and the incumbent service providers planned to launch broadband services to the home. Without fixed-line or wireless broadband access in place, the entertainment industry operates as the oligopoly that it is; when cable networks raise their prices, satellite networks follow suit.

A fundamental substitute became more practical when the Federal Communications Commission ordered local number portability between fixed-line and wireless services in 2003. For many customers, the ability to transfer the local number to a mobile phone gave them the opportunity to eliminate the fixed-line subscription. When customers view the value of one service to be greater than another, adjusted for price differences, they will switch to a substitute service. The pricing of wireless is one example. In 1999, the Yankee Group postulated that wireless service in the United States would displace landlines when the price ratio was three to one [2]. To the customer-centered telecommunications service provider, this ratio is more important than regulations or competitors' promotions. There were examples of fixed-line substitution around the world years before local number portability was mandatory. Local fixed-line service in Europe was never offered at an unlimited price, though the price was lower than wireless when cellular service came to the market. Furthermore, Europe's policy of "calling party pays" encouraged wireless users to leave the device connected and nearby, and more importantly, give out the number to potential callers. Flat-rate pricing for local or long-distance wireless minutes within the United States caused significant substitution for home long-distance services. Customers signed up for wireless plans and committed to hundreds or thousands of bundled minutes to take advantage of low high-volume rates. Once obligated, the customers would consume the minutes by using the wireless phone rather than a nearby landline, which might have offered lower per-minute rates.

The customer-centered telecommunications service provider will be an ardent student of *buyer* power. Buyers in the consumer segment exercise power when they demand higher service quality, more features, and lower prices. Few switching costs between service providers coupled with undifferentiated service cause a high level of churn. Low growth in a segment creates buyer power, as the service providers scramble to gain new business by taking customers away from their competitors.

Buyers in the business segment are most powerful when they represent a high proportion of overall revenues to the service provider. Business buyers are knowledgeable, which gives them negotiating power against service providers. Moreover, they are typically among the most profitable customer segments, and thus are in high demand by service providers. Buyers in highly concentrated industries exert leverage in negotiations and demand a high level of attention during the business relationship. Telecommunications service providers are further hampered because large buyers' of full-featured services can choose to bring the function in-house by purchasing, customizing, and managing the components on their own; for those business customers who already buy the wholesale components, their commodity nature makes them very vulnerable to price competition.

2.2 Marketing Telecommunications Services as a Commodity

Commodities are products and services perceived by customers to be exactly the same as those offered by competitors. A commodity or near-commodity has a high level of substitutability with its competitors. In a commodity marketplace, the major competitive arena for the two offerings would be price. Pricing is straightforward to customers, who can compare rival offerings readily. In most markets, though, pure commodities are rare if they exist at all. A cola in the downtown supermarket is not exactly the same product as the one in the corner store, nor is it the same one to buy at four o'clock in the morning, when only the all-night convenience store is open. Telecommunications services are commodities when the customer is indifferent to competing offerings. A service provider can launch a million-dollar advertising campaign to burn a brand name into the public's consciousness, but if the customer is indifferent to differences between suppliers, the service is a commodity.

Escaping commodity perception and commodity pricing requires the marketer's creativity and commitment. The first step is to identify those customers or types of customer that are receptive to differentiation. Understanding the size of the market and the potential premium price that those customers are willing to pay establishes limits upon how much investment is available to differentiate the service from competing commodities. Typically, the customers that can find value in differentiated services are businesses, as they quantify the benefits that they will gain against the price premium of a differentiated service. Consumers who are receptive to a differentiated service might act on instinct rather than financial analysis, and the premium they are willing to pay might exceed the difference in cost by a significant amount. Equally important is to identify those customers who would never pay a premium for service differentiation. Service providers need to know which of their customers are commodity seekers and which are value seekers so that marketing investment is never addressed to commodity seekers. Customers lured by low price are willing to find the lowest price service and willing to leave as soon as the lowest price is gone. Promotional pricing, bonuses, and advertising are wasted on these short-term customers, who are most often lost before the marketing investment is recovered.

The challenge for the telecommunications service provider is to differentiate its services from those of its rivals or pursue a low-price strategy for undifferentiated services. When differentiating a service is successful, the service will draw customers and market share from competitors or gain a premium price, or both.

Whether one service is different from its competitor is not, in the end, up to the service provider; the difference is in the customer's mind. If the customer sees the alternatives as equal, then the customer will purchase the cheaper of the two. Thus, the service provider who chooses to be the low-cost provider can ignore differentiation strategies and concentrate on price reduction. Service providers that target commodity buyers work on the supply side of the market, while differentiators work with customer demand to offer value not available in competing services. Commodity service providers exercise little control over the prices that they command for their services; the most profitable strategy is a focus on reducing costs.

The low-cost provider can offer both commodity and differentiated services, but the service provider needs to know which one is which. Is a long-distance service with free calling on Sundays, frequent-flyer airline miles, and on-line billing a differentiated offering? It depends whether the customer can find the same package, or one in which the differences mean nothing to that customer, from another service provider.

Choosing to market a service or a line of services within a price-driven market does not preclude developing a brand identity. On the contrary, service providers can create a low-cost brand and enhance it with a discounted image, though a successfully branded commodity no longer operates as a commodity. AT&T and MCI each created a new brand for the low-cost dial-around services that competed with their own subscription long-distance calling plans. Significantly, neither company associated its deeply discounted dial-around brand with its own carefully protected corporate brand.

In another application, AT&T required on-line billing for customers of its lowest cost long-distance subscription services. Commodity seekers understand that on-line billing reduces the service provider's cost. Brands, like commodities, exist only in the mind of the customer. A brand encapsulates the customer's expectations of the branded service. Furthermore, a commodity's brand might very well be the only difference between two competing products. Facilities-based service providers can resell their unbranded bandwidth through branded retailers without identifying themselves to the customers who buy their own branded services at a different price. Whether the wholesaler's price for an otherwise identical offering is higher or lower than the retailer's depends on how well each of the service providers differentiates the service in their own markets.

To be sure, a commodities market for bandwidth is in place, and for the most part, the confluence of supply and demand will determine price. Yet even for wholesalers who choose to participate in selling bulk minutes of use, opportunities abound for differentiation. First, of course, reliability guarantees in the form of service-level agreements change the price profile. So do the quality of the connection and the availability of the bandwidth after an order. Telecommunications service wholesalers can offer an array of services at commensurate prices that are competitive or under the prices offered by competitors. The array itself might serve to add value to the bandwidth; flexibility in these parameters might assist the wholesaler's customer to vary its own services and prices to those customers interested in variable price options. Further differentiation tactics can include network management features, creative financing solutions, and network consulting services.

Furthermore, the commodity service provider can choose market leadership through preemptive price decreases. For most of the 1990s, Sprint's long-distance customers could be confident that their service provider would decrease per-minute prices, package uncomplicated services, and offer promotional bonuses before, or at least soon after, its larger competitors AT&T and MCI. Customers of commodity services in large quantities seek competitive prices, but they also want to benefit from cost reductions without delay from their suppliers. Furthermore, customers who bind themselves to long-term contracts to reduce or at least lock in their costs for the term of the agreement will expect that price reductions will at least match those in the marketplace as soon as they are available.

2.3 Differentiating Telecommunications Services

Opportunities to differentiate services are present in each element of marketing, beginning with the marketing mix of product, price, promotion, and distribution. Though product differentiation comes to mind first as a marketing strategy, some of the most noteworthy—and least costly—marketing successes have sprung from innovations in other aspects of the marketing mix. Indeed, a glance at the mission statements and public documents of too many would-be market leaders in the industry reveals an overemphasis on undeployed, immeasurable or unimportant service features ("we offer end-to-end service and one-stop shopping"). A dearth of innovation or drive for leadership in other aspects of the customer experience leads to insipid and unmeasurable goals ("we want to be our customers' choice for high-quality, high-value telecommunications services"). Alternatively or additionally, service providers might offer attributes of the service that customers fully expect and which do not differentiate the service at all ("customer service available 24 hours a day, 365 days a year!").

Product differentiation can occur in the traditional places, such as the quality of the connection and the availability of the network. But network quality is a better differentiator in the business and wholesale markets than in the consumer market with the possible exception of wireless services. While failures in quality or availability will indeed cause customers to churn, without network management tools, few consumers can discern the service provider's performance against these metrics. Fixed-line services in developed countries boast excellent quality that no longer differentiates one service provider from another. Still, differentiating on the basis of perceived service quality is a combination of network management and customer communications. Sprint managed to sustain an image as low-cost provider and high-quality provider through a program of price leadership, network management, and an advertising campaign that reminded consumers constantly of a pin dropping. The mantle of perceived network quality recently passed to Verizon, according to J.D. Power and Associates, perhaps not coincidentally after its successful and ubiquitous "Can you hear me now?" wireless campaign.

Moreover, service providers can differentiate on the basis of value-added services, including information (such as Internet access for wireless customers, entertainment content, or network management tools), innovative network provisioning features or customized billing options in the retail market, or on-line customer service. Voice over Internet protocol (VoIP) deployment will enable service providers to build innovative and customized applications for both wholesale and retail customers.

Marketers of telecommunications services can directly differentiate their service offerings through creative pricing, including service bundles and structural changes. Customers can select among an array of price points based on the class of service that meets their needs. Ideally, the service provider that innovates price packages will focus on new structures that are not easy for competitors to match, through proprietary information technology or peculiarities of the service provider's market position. AT&T was the first wireless service provider to offer a nationwide rate for minutes, in a program that changed the price structure of the wireless market forever. A latecomer to the wireless market, AT&T was able to offer this price structure because it already operated a nationwide long-distance network. Though other service providers had to match the offer over time, AT&T owned the innovation for some time, signing 200,000 customers in the first 60 days AT&T offered the service.

Promotions are another way to innovate and differentiate. Long-distance service providers commonly use promotions to acquire customers, but promotions are too often an ineffective means to acquire profitable and loyal customers. Advertising and the promotion itself is an investment, such as signing bonuses, discounts, or special prices, and it needs some degree of customer retention simply to recover the original cost. Furthermore, telecommunications service providers who neglect to qualify the type of customer to target for a promotion will draw some customers only for the duration of the promotion. AT&T solved this problem with a bonus program that rewarded loyal customers but not churners. The customer earned small bonuses early in the relationship, and the bonuses increased at intervals, as long as the customer remained loyal.

Distribution channels represent a special means to differentiation, because they embody the service provider's relationship with the customer. The first element that can create differentiation is simply where the customer can buy the service. In the consumer market, distribution outlets include agents, shops and kiosks, service-provider-owned retail outlets, and the Internet. In the first years of long-distance competition, MCI broadened its market share from a tiny upstart to a formidable competitor to AT&T with its Friends and Family campaign. AT&T was unable to respond competitively. First, its near-monopoly market share at the time precluded it from offering large discounts to all of its own customers. Second, AT&T's legacy information systems were unable to cope with the traffic measurement that MCI's new relational databases were able to process. This approach was less a pricing innovation than a promotional one; MCI used its own customers as the distribution network to qualify new customers for its discounted rates.

The service provider whose products are available through all of the channels used by its competitors is commendable, but not differentiated. Service providers achieve competitive advantage only when innovations in marketing add value to the service unmatched by competitors.

2.4 Emerging Marketing Opportunities and Strategies of Industry Leaders

Marketing is the process of defining and communicating the value proposition of a service offering to customers. The *value proposition* is, in the end, the customer's view of the benefits, costs, tradeoffs, and substitutes for a given service offering. For the wireless customer, the value proposition includes the benefits of wireless service, the reliability of the network, the price (especially as compared to competitors), the features and even the chic of the wireless handset, the availability of support, portability, and payment options, at a minimum. All of these elements must compare favorably between service providers, and between having wireless service or doing without it entirely. The value proposition, like the brand, resides only in the mind of the customer. It is the marketer's challenge to shape that perception favorably toward one's own brand at the expense of others.

What are the unexplored opportunities for telecommunications service providers to establish customer value? They exist throughout the marketing mix. Even the product or service itself offers marketing opportunities beyond the technology advances that are normally outside of the purview of the marketer. For example, bundling products is a marketing function that crosses product, pricing, and packaging. Study after study demonstrates that customers are receptive to bundled offerings. The marketer's challenge is to design a service where the bundling improves the services bundled in the same manner as a clock radio creates an object with a value that transcends its parts. The telecommunications service provider that simply bundles two services, only to lower the aggregated price and reduce its own administrative costs, is seeking the least profitable customer, the commodity seeker. The key to successful bundling is to create something new by integrating services. Thus, bundling offers the most innovation only when it alters the product itself.

Pricing deserves considerable attention, not only in its level and structure, but in its timing. Innovations in pricing offer increased profits and sustainable market share only when the telecommunications service provider offers pricing that its competitors cannot match. For example, cost reductions can result in lower prices. But the reseller whose suppliers reduce its per-minute charges will not gain a competitive advantage when it reduces its prices, as its competitors will undoubtedly enjoy the same cost reductions. On the other hand, a service provider that automates a labor-intensive function, such as service provisioning or repair, and uses proprietary technology, will enjoy a cost reduction unavailable to its rivals.

Similarly, the service provider that changes the structure of pricing might or might not benefit with increased revenues or market share. When a wireless provider offers free weekend calling over its underutilized network, its value as innovation lasts only until the remaining facilities-based wireless providers offer the same feature. A more sustainable approach to pricing was AT&T's Digital One Rate, in which the service provider eliminated the roaming charges that kept many wireless customers from using their wireless phones outside of their own service area. AT&T also eliminated the price distinction between local and long-distance minutes and offered low per-minute prices when a customer committed to thousands of minutes per month. AT&T could be generous with long-distance only because it owned a nationwide landline network, differentiating itself from most of the local telephone companies it viewed as its strongest competitors. Furthermore, its closest rival in long-distance calling did not have a wireless service. A relatively new entrant to wireless, AT&T's market share grew dramatically in the months after it offered the new price structure.

Customers of technology are accustomed to price reductions and service upgrades, and the savvy telecommunications marketer will develop repricing strategies that engender loyalty. Sprint established a reputation as a long-distance price leader, both in structure and level. First to promote price innovations to its customers, delayed actions by competitors did not tempt Sprint's customers to churn for lower prices. Sprint enjoyed high ratings on both its network quality and customer service, which led to customer loyalty.

The most fertile ground for technological innovation in marketing is in distribution, from network provisioning, to service quality management, to customer care. Telecommunications service providers have only begun to tap information technologies in these areas, and the introduction of Internet protocol telephony, or its descendents, will provide innovation as yet unimagined in the area of customer care.

References

[1] Porter, Michael E., *Competitive Strategy: Techniques for Analyzing Industries and Competitors*, New York: Simon & Schuster, 1998.

[2] Felps, Bruce, "Study Says Wireless to Challenge Landline," *Wireless Week*, Vol. 5, No. 2, 1999.

3

The Marketing Plan for a Customer-Centered Telecommunications Enterprise

3.1 The Marketing Planning Process

Three baseball umpires are debating their job descriptions, if rather ungrammatically. The first umpire states, "Some of them is strikes, and some of them is balls, and I calls them as they is." The second, slightly more enlightened umpire says, "Some of them is strikes, and some of them is balls, and I calls them as I sees them." The third umpire says, "Some of them is strikes, and some of them is balls, but they ain't nuthin' until I calls them." The customer is the third umpire, and the telecommunications services marketer has to concede that the customer decides, in the end, what the service provider is offering.

This marks one significant difference between the business plan and the marketing plan. The business plan, while acknowledging the demands of the marketplace, focuses primarily on making the service portfolio a reality. The marketing plan's function is to place the service portfolio in the mind of the customer, working within the attributes and weaknesses of the service portfolio. Naturally, these two processes are interdependent and iterative. But the business planner can transform the entire enterprise, and the marketer transforms the customer's perceptions, to position the service portfolio favorably against competitors and substitutes.

The marketing plan offers several advantages to the enterprise. The first is that it is indeed a map, not a brochure. The qualities that make the best marketing professionals successful—persuasiveness, big-picture thinking, enthusiasm that can go beyond the facts of the case—are less useful in providing directives to the rest of the enterprise. Those in the back office who most effectively execute, rather than visualize, marketing strategies are structured, efficient, and analytical. The documentation and direction of a marketing plan keep the enterprise moving in an agreed direction. Furthermore, it forces that quintessential marketer to focus the marketing vision, ratify it with supporting calculations, and commit to its success as written.

All the same, the marketing plan does serve to rally the enterprise with team spirit. Most employees in every corporate and operational function are thrilled to be included in the marketing team, and each employee's contribution can increase revenues, lose customers, or affect the service provider's reputation. Indeed, as one marketing executive once said, "Everyone's in marketing."

The marketing plan commits all of its participants to measurable results within the time horizon of the plan. While a marketing plan might include milestones toward a long-range goal, the marketing plan's objectives should be achievable within a single year. A year is enough time to see if the marketing vision is off-course, and the telecommunications business environment changes so rapidly that planning many specific accomplishments for a longer horizon is impractical.

Lastly, the formal marketing plan that gains the approval of senior management places responsibility for its completion upon all responsible parties. With a financial analysis of investment and results, management agrees to fund the programs outlined in the plan. The marketing organization agrees to deliver customers, which feeds the operational budgeting process. Ideally, the marketing planning process includes the participation of staff members representing a broad range of divisions, who can summarize customer experiences and potential avenues of customer acquisition to open new markets. Accordingly, the marketing plan serves as a contract, an agreement among all involved parties, to achieve harmonious goals to meet the short and long-term marketing goals of the enterprise.

The first step in marketing planning is to review the previous plan, if there is one. For all the attention on very specific business plans, very few enterprises bother to evaluate the previous plan's results in the context of the planning process. The value of this process goes beyond whether the enterprise performed to expectations. It evaluates the entire marketing planning process: whether the assumptions were valid, what divisions did not participate in planning and might have improved the plan or results if they had, and where better planning might have seen opportunities on time to exploit them.

In the 1990s, when home Internet access was moving from the early adopters to an early majority, several of the incumbent U.S. local service providers saw an opportunity to sell second lines to their residential subscribers. Though the offering of second lines, prior to widespread DSL deployment, only offered narrowband access, customers sought to free up the voice line from overuse by the Internet addict in the house. At the time, competition in the local network, especially in the consumer market, was only beginning. Up to that time, local service consumer market planning had been driven by operations planning, which in turn had relied on demographic growth estimates, primarily the plans of residential developers, instead of forecasting demand of actual customers. To the incumbents' credit, aggressively marketing second lines was indeed a real step towards competitive marketing. But company business processes did not readily accommodate marketers driving infrastructure buildout, and capital investment planning did not anticipate the demand for second lines for Internet access. Delays, formerly unheard of in an industry that historically managed its technology introductions slowly and deliberately, occurred with such regularity that state regulatory bodies needed to intervene.

Moreover, the shift from voice to Internet access on lines that were not enhanced with broadband capabilities changed the nature of traffic when new lines were in operation. By 1998, a typical on-line session lasted a half hour, ten times longer than the typical three-minute voice call [1]. The marketing planning process requires historical and projected data from virtually every division in the enterprise to feed a reasoned projection of costs and activities that generate the revenues projected by the plan.

Armed with the revisions to the previous planning process, senior marketing executives establish the planning team. The customer-focused marketing plan requires team members who represent the customer perspective. This is more difficult than it seems, especially because few service providers have the luxury of including actual customers on the planning team. There are employee surrogates for customers who can put aside company loyalty on behalf of the planning process: large-customer account managers, customer-care middle managers, and employees who are affiliated with outside user groups. These employees spend their working hours listening to customer complaints, unmet needs, and compliments. On an event-by-event basis, their normal work routine might address each of these items. The marketing planning process is the place to study them in the aggregate, or even anecdotally. Individual victories or defeats can harbor insights for future marketing efforts.

3.2 Directing the Marketing Plan to the Customer

Any basic marketing textbook or a cursory search on the Internet will unearth dozens of marketing plan elements. The customer-centered marketing plan focuses each of the elements on shaping the customer's perception.

Because the marketing plan should be used, not shelved, the shorter it is, the better. Its impact and objectives should be limited to a single fiscal year. To the degree possible, lengthy explanations of new services, or subsidiary financial statements, or administrative matters that do not assign a task to an individual, should merit only a footnote reference to more information in another place or a hyperlink in an electronic document.

The ideal *executive summary* does not exceed a single page or two made up mostly of white space. This précis of the rest of the marketing plan is the paper equivalent of the fabled 15-second elevator ride with the CEO that a person gets once in a lifetime. If a marketing plan cannot abide by the constraints of a half-dozen bullet points and very clear commitments, then the planning process that arrived at that summary was incomplete. After all, if the executive or board of directors scanning the plan cannot get through the summary, the plan is unlikely to win funding. Making the executive summary center on the customer is a fine way to eliminate text that might better be situated on the company's business plan.

The *situation analysis* section provides the backdrop for the strategies and tactics to follow. Though many of its components do not relate directly to the customer, they do relate indirectly, and the exercise of introducing the customer to the thought process is a useful one. For example, the state of regulation is still an important component of the fixed-line service provider's situation. The residual regulation of incumbent service providers and the tariff-dictated resale rates of unbundled elements affect nearly all of the local wireline service providers in the United States. From the marketing point of view, all that matters is the customer. Are there substitutes for the regulated service on offer, and have recent developments changed the price or utility of substitutes in the view of the customer?

This section also includes a statement about the present market position of the service portfolio, including its strengths and weaknesses, in contrast to the services offered by competitors, and their strengths and weaknesses. A market position only exists in relation to a competitor; if a close competitor is strong on low cost and low price, then one's own offering cannot be strong on price. On the other hand, if prices are high enough among all competitors in the marketplace, then a customer's possible substitute for the service in question is simply not having it at all. Different products in the portfolio might have different attributes, and might indeed be substitutes for each other. One metric that belongs in the situation analysis section is market share, the ratio between a company's revenues and the revenues garnered by the total market.

The situation analysis also assesses the customer situation, which might include the important purchase criteria used by various market segments. These criteria might vary across segments or across services. A slow economy might suppress spending on new wireless devices, but it might also increase business bandwidth in lieu of costly travel. Most noteworthy are changes to customer buying patterns that might signify trends that present threats to existing revenue streams or new opportunities. A severe slowdown in a micro-economy that represents high telecommunications volume, such as the hospitality industry, might threaten service providers whose customer base is overrepresented by the business segment.

Account managers who are close to a vertical market might offer guidance to increase market presence in the industry segment. An important industry segment might warrant its own service portfolio, or a dedicated customer care team, or proprietary software that creates value and differentiation from competitive offerings. The marketing planning process can create such infrastructure, along with a budget and a plan of operations within the segment. Demographics, while no longer the main element of marketing planning, remains of interest in the marketing plan, which seeks novel segmentation strategies and unique blends of services and pricing to appeal to a variety of consumer profiles.

The situation analysis is also the opportunity to assess the state of technology, including identifying technologies that have matured and reached the end of the life cycle, and examining underutilized or unexplored technologies that might provide new service lines, bundling opportunities, or substitutes for the present portfolio. The planning team can recognize fading technologies by the migration of users to newer models, by the slowdown in adoption during the phase when only laggards finally join the late majority, and by deep revenue slides as the customer's perceived value falters. The marketing plan can address emerging technologies that are not yet in the marketplace by establishing incubators, focus groups of innovators, or develop a research plan to study ongoing use of the new technologies in more advanced markets around the world. By no means is the technology assessment a technical description. Its purpose is to evaluate existing and potential markets for technologies available for deployment within the time horizon of the marketing plan.

After the analysis of the surrounding environment, the market plan should address the service provider's strategic approach to the marketplace. In their book, *The Discipline of Market Leaders* [2], Michael Treacy and Fred Wiersema identify three marketplace strategies that they call "value disciplines." The first, operational excellence, offers services at the least price and at minimal inconvenience. The second, product leadership, offers leading edge or superior quality telecommunications services to its demanding market. The third, customer intimacy, seeks to meet the particular needs of its target market segments. For the most part, these disciplines are mutually exclusive approaches. The service provider whose focus is low price will not be able to offer the unassailable quality demanded by the customer of product leadership, nor will the price be the key criterion for the customer who seeks a customized solution and a high level of attention. Thus, the service provider must make a commitment to excel in one approach and invest in the marketing and operational infrastructure to succeed in a crowded marketplace.

For the telecommunications services marketer, the fundamental approach should be part of the corporate philosophy before the marketing planning process begins. If it is not already clear from the previous business plan and corporate mission statements, marketing management needs to obtain direction from senior management and plan marketing strategies and tactics accordingly.

Marketing plan strategies are not tasks, but they should generate specific objectives and the tactics to fulfill the objectives. Thus a strategy might include a program to create excitement about a new service, and individual tactics would support the strategy through market research to identify segments of likely demand for the service, awareness advertising campaigns, signup promotions, collateral materials, and publicity.

One section of the marketing plan should focus on *objectives*, the proposed scorecard for the plan. Each objective can correlate with a specific tactic within the larger framework of a marketing strategy. Some objectives might cross strategies, such as a commitment to reduce churn by a percentage point in a particular service line. The objectives contain the scorekeeping, so they are well suited for presentation in the executive summary of the marketing plan, if only in bullet-point form. Somewhere deeper in the document, objectives carry other attributes that give them meaning. For operational parameters such as telemarketing close rates, these attributes include the metric to meet, and in some cases, the measuring device used to calculate the metric. All objectives should include a completion date and a responsible manager. If the objective will require significant investment or additional budget for labor, expenses or capital assets, the marketing plan should present the amount and a breakeven analysis.

The customer-centered marketing plan includes the marketing mix of product, promotion, pricing, and distribution with the customer perspective throughout. In practice, for example, the section that describes a new pricing structure substantiates its proposal with data from focus groups or test markets, research on market segments that might be especially responsive to proposed service bundles, or survey results that indicate an unmet need in the marketplace. The customer-centered marketing plan tests each objective against the customer. If an objective does not address a customer need, its inclusion in the plan should be reconsidered.

3.3 Differentiation Strategies

A focus on customers does not ignore competitors, because the battle for market share takes place inside the customer's mind. Rather, a customer-centered strategy recognizes that competitors endeavor to stand between you and your customer, and that your goal is to keep your existing customers while intercepting new customers from your competitors. Al Reis and Jack Trout likened competition to warfare by borrowing the principles of an 1832 book called *On War* that had been written by a retired Prussian general. The resulting treatise, *Marketing Warfare*, identified strategies that used proven military strategies to explain marketing successes [3].

The principle of force states that the competitor with the greater "troop strength," all else being equal, will be victorious. Indeed, in the contest for industry leadership in any year, the way to bet is to pick the leader from the year before. Size creates force, whether the sales force, or the mind position of the leading brand, or simple inertia in cases of only average customer satisfaction. Thus AT&T did not end its generic "Reach Out and Touch Someone" campaign as soon as it lost its long-distance monopoly. A service provider that owns 90% or more of a market can expand the market and be sure to get nearly all of the new revenue.

Nonetheless, service providers just entering the industry or holding very small positions in the overall market cannot fight the industry leaders in all places at once. Level 3 Communications entered the wholesale transmission services business in 1998. Never expecting to serve the retail market, the company focused on building an Internet protocol-based network unencumbered by legacy facilities. Within a few years, Level 3 held a 32% share in the wholesale dialup services market, the second largest provider after MCI's UUNet [4]. By 2004, its customer base included nine of the world's top 10 telecommunications carriers, nine of the 10 largest European carriers, and five of the top 6 U.S. Internet service providers. Though Level 3's infrastructure is among the largest communications and Internet backbone networks in the world, it has stayed in the managed networks marketplace, serving the largest Internet service providers and corporate customers. Its focus of resources on a narrow and underserved segment enabled it to grow quickly and establish a strong brand within a market otherwise held by enormous but diversified competitors.

Ries and Trout identify a "strategic square" for marketers, including a defensive strategy for the market leader, an offensive marketing strategy for a strong second leader, a flanking strategy for those that do not have the resources to take over the market, and guerilla marketing for small firms whose only opportunity to dominate a market is to dominate a small segment.

Applying this concept to the wireless market, at the end of 2003, wireless market share in the United States was split among the six primary service providers, with Verizon in the lead with about one-quarter of the market, and AT&T and Cingular in battle for second place, each with about 15% of the market. Thus their combination would reposition the service provider from a second-place tie to a first-place leadership. Sprint, Nextel, and T-Mobile followed, and the remainder of the market consisted of small service providers. These market shares, when displayed in pie chart form in Figure 3.1, demonstrate the lead that Verizon commanded in the wireless market at the time.

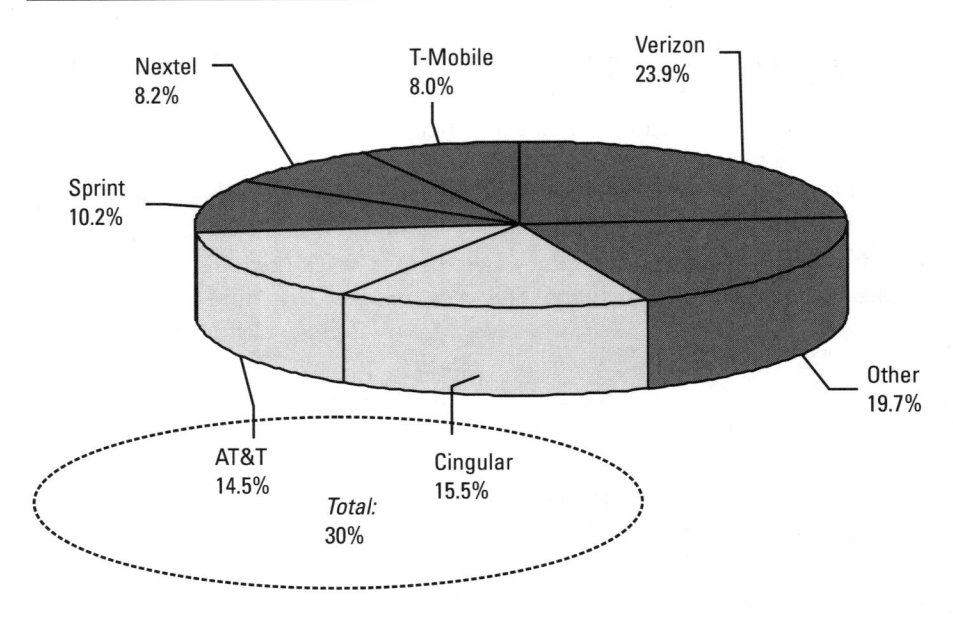

Figure 3.1 Wireless U.S. market share. (*After:* [5].)

Pie chart segments: Nextel 8.2%, T-Mobile 8.0%, Verizon 23.9%, Other 19.7%, Cingular 15.5%, AT&T 14.5%, Sprint 10.2%. AT&T and Cingular Total: 30%.

Thus, Verizon could operate defensively, which gave it an advantage over its rivals in retaining and even increasing its share. The adage "nobody ever got fired for buying IBM" demonstrates the power that market leadership affords for the undecided buyer. Cingular and AT&T were close rivals to Verizon, but they were not the leader, and each only held about three-fifths the market share owned by Verizon. Thus, their strategies would not succeed if they applied marketing resources head-on against Verizon (though they could compete more evenly against each other). AT&T and Cingular needed to apply offensive strategies against weaknesses in Verizon's position or frontal assaults against each other. Yet each of these service providers lost market share to Verizon between 2002 and 2003. This often occurs among second-place contenders who meet the leader in the marketplace without an offense strategy. AT&T and Cingular could not beat Verizon with similar plans and services without exploiting weaknesses in its line. When the companies combine, a frontal assault strategy could be effective.

Nextel and T-Mobile both experienced market share growth in the same period in which market shares for AT&T and Cingular were decreasing. Nextel accomplished its impressive gain through its proprietary push-to-talk technology. T-Mobile focused on the high-end customer segment with its GSM network and pioneering data services. These are flanking strategies against the industry giants, and both were successful for these smaller service providers. But both of these innovative technologies are susceptible in the long term to imitation from other service providers, including the market leader. Both Verizon and Sprint upgraded their networks to offer push-to-talk, though Nextel's first-mover advantage gave it an excellent head start and an ability to be the standard-bearer. Similarly, technology in later generations of wireless technology will merge the industry protocols, and T-Mobile will need to develop technologies or other differentiators to sustain its status in the wireless market. In support of sustaining its growth, the service provider has leveraged its market position using partnerships with larger wireless carriers and innovating in Wi-Fi access in the United States. Sprint's future market position has been challenged by local number portability and a nebulous market strategy. According to Strategy Analytics, nearly 40% of Sprint customers expressed an interest in switching service providers if number portability was available, the highest percentage of customers favorable to portability for all major wireless service providers. In the same survey, Verizon would be the most popular destination for churners, validating a defense strategy for the market leader.

For the service provider that falls in the bottom 20% of the market, the appropriate strategy is guerilla marketing. Omnipoint, a U.S.-based GSM service provider and predecessor to T-Mobile, created a tiny niche among frequent international travelers. GSM, while in limited availability in the United States, was the protocol of choice in all of Europe and most of the rest of the world. Omnipoint developed roaming agreements with other countries, creating per-minute pricing that was not very different than roaming prices between U.S. cities. More importantly, Omnipoint's overseas roaming rates were a fraction of the prices for non-GSM providers such as AT&T. Frequent overseas travelers found Omnipoint's international rates lower than hotels or calling cards, and the convenience priceless. Beyond the traveler's ability to make calls from anyplace comfortable, the Omnipoint customer could use any dual band or tri-band telephone to find a roaming network. Omnipoint targeted its advertising by presenting collateral materials in places that its prospects congregated, such as airport executive lounges and hotel lobbies. For a small service provider with a very limited marketing budget, extremely targeted marketing and high effectiveness are critical. After Omnipoint's merger with VoiceStream and then its acquisition by Germany's T-Mobile, Omnipoint's pricing structure moved in line with its nearest competitors. To stem potential churn from its original customers for whom roaming internationally was the singular reason for subscribing, a grandfather strategy kept them from switching.

One effective strategy for the industry leader is to attack oneself. Entrepreneurs created a large dial-around segment in the early 1990s largely through a flanking strategy. Targeting low-cost seekers in the undifferentiated long-distance market attracted 15% of consumer callers [6]. In 1997, MCI joined with an unbranded alternative called Telecom USA and captured a leadership position while legitimizing the market with national advertising campaigns.

Eventually, AT&T weighed in with its own offering, thus attacking its own leadership position with an unbranded alternative. Leaders can also copy competitive actions before their own market share erodes in favor of the innovations of others. Verizon was the first wireless service provider to launch push-to-talk technology in response to the Nextel threat in an attempt to slow Nextel's momentum.

For those attempting an offensive marketing strategy, the best attack point is a weakness inherent in the leader's strength. After all, given time and investment, leaders can correct market weaknesses once competitors identify them, as Verizon did with its implementation of the missing push-to-talk capability. On the other hand, the industry leader cannot eliminate its strength to meet a competitor's strategy. Sprint was the first long-distance provider to boast an all-fiber network, and it took every marketing opportunity it had to boast about it. Even the advertisements that did not mention the high quality of its network managed to drop a pin with a little pinging sound, just to remind consumers of its mind position. This approach was a direct assault on its closest competitor, MCI, whose network was thought to be cheaper but inferior to AT&T's. But the magic of this campaign was that even AT&T could not retaliate with its own fiber optic network, thus lending credibility and conceding defeat to Sprint's newer and much more targeted investment. Furthermore, it would have been futile for AT&T to disparage the notion of high-quality fiber, as its own plans included a transformation to fiber in the future. The same marketplace battle continues as the newest entrants to the marketplace boast of their VoIP networks and will offer features unavailable or costly to buy in traditional voice networks.

Flanking marketing strategies target those customers that are overlooked and underserved. Most new telecommunications markets emerge from technology advances, such as camera phones or data access in the wireless market, or broadband services over phone lines or cable. But marketers can create new services and new segments simply through the marketing mix. The prepaid calling-card services market provides an excellent example of a sector that emerged when entrepreneurs simply repackaged and distributed resold telecommunications services. Most vendors of prepaid landline services do not own networks or support a large labor force. Their business model focuses on low per-minute international rates and carefully maintained distribution outlets. Prepaid service providers focus on market segments that are not targeted by industry leaders. Customers of prepaid long-distance services include students, immigrants, and consumers without bank accounts, customers that are ignored or actively avoided by service providers seeking subscribers. A 2002 estimate by research firm Atlantic-ACM projected growth in U.S. prepaid calling-card services revenue to slow from the compounded annual 25.4% growth rate experienced over the previous 7 years to a more sustainable but still robust 9.7%, reaching

$6.4 billion by 2008, a 73% increase in a 5-year period. The industry, with a life of its own, separate from its long-distance roots, now includes prepaid wireless, local service, and Internet access.

A bond trader whose dream was to build a worldwide fiber optic network founded Global Crossing in the 1980s. Its $15 billion 100,000-mile network reached 85% of the world's telecommunications markets [7]. The company strategy was guerilla warfare: serve a tiny but very profitable market segment by being the carrier's transatlantic carrier. This strategy attacks the larger international networks outside of their main markets, but uses their customers against them. AT&T's competitors without their own international networks would be happy to buy bandwidth on a state-of-the-art network from any reputable service provider that wasn't trying to steal its retail customers. Subcontractors like Tyco built the undersea cable for Global Crossing, while Global Crossing stuck to its skills in funding the efforts. In 1999, the company acquired a cable installation and maintenance division from Cable & Wireless PLC. As a guerilla marketer, Global Crossing might have succeeded as the carrier's carrier it promised to be, with enormous bandwidth capacity at low, fiber-driven prices. But the company's acquisitions overreached its marketing strategy and its size. First, it tried and failed to acquire incumbent local service provider US West, losing the battle to another professed wholesaler Qwest, who inherited not only the victory but also the strategic conflict of interest. Global Crossing more successfully acquired a different incumbent local service provider, Frontier Corporation. A series of difficulties led Global Crossing to bankruptcy and redemption. Overcapacity in the market for bandwidth reduced prices for all service providers. Technology companies were losing the stock market valuations that had enabled them to acquire the venerable service providers such as Frontier Corporation. But Global Crossing alone was responsible for the disconnect from its guerilla positioning that might have saved it from bankruptcy and even the creative accounting that led to scandal. It acted like a full-service industry leader. The company contributed more than $3 million to candidates in the 2000 U.S. elections, acting like the entrenched regulated service providers. The company made millions of dollars of charitable contributions, a noble undertaking but one not normally associated with guerilla marketing tactics. The executives took salaries, stock options, loans and bonuses that rivaled those in the largest U.S. corporations. Guerilla marketers need to stay lean, to enter and exit markets flexibly, as the climate warrants. Global Crossing left its strategy behind as if it had become the market leader.

It is never surprising to see an entrepreneur whose company is acquired leave the corporate world soon thereafter. The skills that define the guerilla marketer simply do not translate well to the corporate world, and the entrepreneur is often happy to leave the stiff, structured world of corporate life. Guerilla marketing works when the service provider is committed to the vision of remaining an outsider, finding a market that stays well away from the rest of the industry, and standing firm against the temptations of the executive suite.

References

[1] Machlin, Robert, "The Internet: Redefining the WAN," *Telecommunications Americas Edition*, Vol. 32, No. 1, 1998, pp. 48–49.

[2] Treacy, Michael, and Fred D.Wiersema, *The Discipline of Market Leaders: Choose Your Customers, Narrow Your Focus, Dominate Your Market*, Reading, MA: Addison Wesley Longman, 1995.

[3] Ries, Al, and Jack Trout, *Marketing Warfare*, New York: McGraw-Hill, 1997.

[4] Long, Josh, "Level 3: Plenty of Room for Growth in Narrowband," *PHONE+*, Vol. 17, No. 4, 2003.

[5] Technology Business Research, "Network Business Quarterly (NBQ) U.S. Mobile Operators 3Q03 Benchmark and Metrics," press release, 2003.

[6] Henderson, Khali, "New Tricks of the 10-10-XXX Trade," *PHONE+*, Vol. 14, No. 10, 2000.

[7] Palazzo, Anthony, " 'Monster' Telecom Network Draining Oceans of Capital (The Rise & Fall of Global Crossing)," *Los Angeles Business Journal*, Vol. 24, No. 7, 2002, pp. 22–23.

4
Market Research

4.1 Data Mining and Customer Profiling

One might expect that the customer-focused market researcher would opt to go straight to the source, the customer, for market research information. While customer surveys and focus groups have their place, they are no longer the only source of knowledge about customer desires. Asking customers what they want is no longer the best indicator of customer needs. Indeed, many customers simply do not know what they want, especially consumers looking at technology products offering features that innovate rather than automate. As consumers normally do not cling to a financial breakeven analysis, buying parameters are softer and more difficult to communicate than the purchase criteria of business buyers. Consumers cannot imagine the benefits of new services, and service providers simply need to tell them, or show them, the new ways that services are important to them.

For example, service providers often provide free introductory pricing for services whose benefits are difficult to explain. ISPs in the United States, where monthly subscription fees for unlimited use stayed at about $25, gave their new subscribers a free month to access the Internet, try e-mail, and create their own demand for a continued subscription. Wireless provider Vodafone, as one example, introduced GPRS data services in Malta with unlimited free access for its contract customers while awaiting the development of its pricing. Many Internet service providers in Europe, where the local telecommunications access providers metered the local connection time, needed to offer free Internet access and gain their revenues through advertising or commissions from the vendors reaching their subscribers.

Another approach is to educate consumers through advertising. Wireless service provider T-Mobile introduced its camera phone with a commercial that targeted young consumers, who were likely early adopters of the feature. Two young men are shopping for a gift for the girlfriend of one of them. When they narrow the candidates down, the unattached of the two secretly calls his friend's girlfriend. He shows her a surreptitious photograph and she signals her approval. He pretends to decide to select that dress on his own, and it is a win-win-win situation: great gift, great boyfriend, great best friend. While the tone of the ad was not very different than an ordinary situation comedy, viewers did not realize that they were in technology adoption school.

While both of these approaches have been used for decades, their newness stems from the widespread use of data mining, also called data warehousing. It was no coincidence that T-Mobile targeted a particular segment with its buying-a-dress campaign, a segment that had no doubt been identified by detailed analysis of usage and adoption patterns within the service provider's own database. Similarly, free trials educate the prospective customer and encourage experimentation; moreover, they qualify the prospects for potential targeting. The user who consumes a high volume of services, whatever they are, during a trial period, is an excellent prospect for a sale. Moreover, whatever profile is common to high-volume users during a trial period will constitute a market segment worth targeting.

Data mining is valuable for optimizing operations, organizational effectiveness, and technology management. Through the discipline called analytics, data mining can leverage the marketing function with dramatic results. Databases offer the most effective window into actual customer behavior, analysis that is not shaped by anecdotal evidence, flawed memory, or subjectivity. Furthermore, data mining can test and uncover valuable correlations from the data itself, rather than from a specific request.

Telecommunications service providers who operate in intensely competitive markets need to go beyond intuition and anecdotal marketing strategies, and technology presents them with the means to do so. Data mining, the process of culling strategic diamonds from customer data, is the confluence of three related technological advances: massive computing power, enormous historical customer records, and sophisticated mathematical algorithms and heuristics.

Competition in data-mining software development has reduced the overall costs of establishing a research operation, especially the startup costs. When pioneering service providers launched their initial operations, the hardest part was finding, sorting, and validating the data pulled from remote regions of the organization. Startup costs of finding and preparing the data fell from 85% of the work effort in the early 1990s to only 15% a decade later [1].

Telecommunications service providers are excellent candidates for data mining. Few industries can map the customer's product or service usage with as much detail and accuracy. The technology decisions center less about finding data to collect and more about finding data to discard.

Data mining can improve marketing efforts in dozens of ways. The research can profile prospective customers, directing the service provider to high-profit market segments and identifying likely churners to avoid investment. Service providers can monitor usage patterns and create service bundles or targeted promotions to increase usage and gain customer loyalty. A review of a single customer's record provides insight that enables the service provider to conduct one-to-one marketing, devising an individualized plan unassailable by competitors that might have equivalent software but lack the data to compete.

The software interrogates a mammoth amount of data to construct a predictive model. The model can describe the likely actions of existing or unknown customers just from the actions of the population about which information is available. Software developers SAS, IBM, and others describe the many service providers who mine their customer databases for competitive advantages. Indeed, many of the world's largest service providers are listed as customers on multiple business intelligence developer sites. Hong Kong wireless service provider Hutchison Telecom used data mining to develop customer profiles and target marketing initiatives to the highest value prospects. South Africa's MTN used data modeling of historical customer records for forecasting handset sales. While the justification for the effort was to reduce stranded inventory, the analysis uncovered unseen trends, enabling the wireless service provider to anticipate market needs and meet them profitably. Italy's Telecom Italia Mobile (TIM) used data mining to reduce churn, predicting likely churners and targeting them with retention offers. Processing 100 million call records per day, TIM culled a 3-terabyte database for marketing and customer retention opportunities. Australia's Telstra Mobile used data-mining technology in preparation for number portability, determining which customers were most likely to churn and for what reason. Targeting vulnerable customers before they make the decision to churn is much more cost-effective than trying to win them back after they are gone. British Telecom built a model that associated each customer profile with their likelihood to purchase and their comparative value once they were customers. Armed with this information, the service provider was able to develop targeted marketing campaigns for the most valuable segments. Using predictive modeling, Canada's SaskTel was able to target customers and improve its direct marketing success rate to 15%, three times its performance against random lists of leads. The service provider's ability to identify customers who respond well to cross-selling means enabled SaskTel to produce campaigns quickly as opportunities arose, without intruding its efforts on customers who were unlikely to buy.

One unexpected repercussion of the gold rush to mining applications is that service providers can develop a false confidence that their investment gives them unique competitive advantage. While the smallest service providers probably are not using sophisticated tools, the largest telecommunications companies worldwide are indeed fully mined. Furthermore, the newest software, having passed through several generations of version management, is so user-friendly that nonmathematicians can create results without understanding their significance.

Data-mining technology has spawned the legend of "beer and diapers," a parable that has since taken on a life of its own. While its authenticity is controversial, it demonstrates the power of data mining; the service provider can poll the data, discover nonintuitive relationships, and adjust its marketing tactics. Data-mining software developer Teredata takes credit for discovering the affinity, recounting it in an article retrievable on its Web site. Apparently, men purchase beer and diapers on Thursday evening in preparation for a tranquil weekend. In response, the legend continues, retailers moved their beer displays close to the diaper aisle and charged full price on Thursday evenings. But the affinities revealed by data mining do not stop at retail. In an appealing metaphor, credit applicants who filled out the application forms in pencil were more likely to default than those who worked in ink.

How does data mining assist the service provider? A database might unearth cross-selling and up-selling opportunities. By evaluating the characteristics of customers who began with a small account and bought more services later, the software can build a profile. While the answers might not be so arcane as ink on a credit application, the service provider might find that customers who began their service on-line, or who pay on time, or who make international calls, are most likely to add services to the market basket. This discovery might lead the service provider to offer promotions to on-line customers, bonuses such as a trial period for an enhanced service for on-time payment, or frequent flyer miles as an incentive to upgrade to a new technology.

Data mining enables the service provider to look back and look forward in an effort to reduce churn. The system can generate profiles of customers who leave and their reason for leaving, if a compatible customer relationship management (CRM) system is also in place. The system then reviews the base of existing customers to find individuals who display the likely characteristics of someone about to churn. In this case, the service provider can also evaluate the life cycle value of the customer and decide whether retention is an advantage.

4.2 Primary and Secondary Market Research

Depending on the service provider's objectives and its budget, it can undertake either primary or secondary research, or both. Primary research entails collecting data, choosing analytical tools, and conducting one's own analysis of the data. Secondary research includes any research produced by outside sources and offered to interested parties. Some secondary research is free, such as government reports, chamber of commerce publications, trade association press, and the press releases of market research firms. Other secondary research costs thousands of dollars, such as the weighty and detailed reports that those press releases from market researchers announce.

Users of secondary research can control their expenditures to meet almost any budget. Another advantage of secondary research, beyond price, is that many of the service provider's research needs do not require the intensity or accuracy of a primary research study. The approach to a new market requires some primary research analysis, but a preliminary review of secondary research might be enough to convince management that the market is not a desirable one. A broadband reseller considering a new geographical market for DSL might discover that DSL subscriptions were not expected to materialize in that localized demographic market because of a low penetration of personal computers, or that growth predicted for the service had not materialized due to high prices, or that other resellers with access to the same bullish forecasts that piqued the reseller's interest have diluted the opportunity. A reseller late in arrival to the long-distance market might learn that revenues per minute were down, or that traffic growth had already detoured measurably to wireless or the emerging VoIP sector.

Secondary research has limitations, and the astute marketer will need to decide whether secondary research meets the standard for making significant marketing decisions. While Internet searches can offer more data than a marketer can use, the search engines do not rate the quality of information they retrieve. A market projection reported by a premier market research firm that specializes in the telecommunications segment, or repeated in a trade publication, is in most cases more reliable than an offhand comment from a newsgroup participant. Indeed, all predictions, whether from five-figure research reports or community newsletter articles, are simply somebody's best guess. Even the blue-chip research companies do not publish scorecards comparing their 5-year market predictions with historical results. Another limitation of secondary research is its availability to anyone willing to obtain it, whether it is available for download on the Internet or available for purchase. The service provider seeking competitive advantage cannot count on competitors to overlook valuable market research. In fact, the largest service providers obtain the leading research simply to maintain competitive parity. Moreover, some publications, including trade journals published by industry lobbying groups, maintain an editorial bias. While reputable publications never knowingly publish inaccurate forecasts or data, they might not include studies that do not support their editorial stance. Furthermore, some necessary research is simply too detailed to be available publicly. A wireless service provider that wants to know how much young adults would pay for a certain ringtone versus another cannot hope to learn the answer from a Web search, or even an industry report. Thus, secondary research is necessary, but not sufficient, for gauging the customer's needs, willingness to pay, satisfaction levels, or future buying patterns.

Occasionally, a market research firm will conduct primary research on a new client's behalf at no charge, especially if they can then sell the results to others. While this does not keep the conclusions private, it keeps the cost low or even free. Moreover, trade associations often conduct surveys of their members and publish the results. Such efforts, especially when member identities are kept secret, will provide valuable market positioning information. Consumer market research firms such as J.D. Power and Associates publish satisfaction metrics for the leading service providers in the largest markets, available at no charge on their Web sites.

Primary research can take advantage of the data archives that the service provider already owns. Data mining, for example, is primary research at its most effective. Another advantage of primary research is the degree to which it can segment and granulate market information. The disadvantage of conducting primary research is its cost. Even in-house data mining is a significant investment. Hiring a research firm costs many multiples of the price of a thick market study. But primary research, especially research that is limited in scope, tightly managed, and with clear objectives that do not change during the process, will pay for itself in market knowledge.

Customers provide a fine source of research without the need for a formal study. With a few changes to normal processes, the service provider can develop a research program that is not quite scientific but useful anyway. For example, each time a customer calls to terminate service, or to eliminate a profitable feature, or even to add a feature, the service representative can document the reason. Without a statistical algorithm, and using only a spreadsheet, the marketing analyst can at least identify the three primary reasons that customers change their minds about their needs. With a bit more sophistication, service providers can conduct surveys and subject the results to statistical analysis with professional assistance.

Interviews with representatives of customer groups and participation at customer group meetings are cost-effective methods of primary research. The sales representative representing a vertical market should at least attend the primary industry meeting each year. First of all, industry specialization requires industry knowledge. Second, the sales representative can identify customer concerns that a customized telecommunications solution might solve. Third, a few well-directed questions to participants outside of seminars might reveal the customer's concerns about their telecommunications services or their present provider.

If customer surveys are possible, how does the telecommunications service provider query the customer about services that never existed before? Regulated local service providers sometimes underestimated demand for the computer-based services like call forwarding and three-way calling when they used standard survey methodology to query the target population. Surveys of willingness to pay were important, because electronic services were high in fixed costs and low on variable operating costs. A broad estimate of price could almost take the cost of the computer and divide it by the number of customers. In fact, as sophisticated marketing software did not exist—nor was it in great demand in a monopoly environment—there was no way to target the population of early adopters. Instead, cheerful surveyors called customers or stopped them in shopping centers and asked, "If you could pay to forward your telephone to another number, what would that be worth to you?" Having no recollection of any situation in which they wished they had call forwarding, customers regularly said that they had no interest. Ironically, those who finally signed on for enhanced services paid well more than they would have if the adoption data had been more accurate.

Other primary research options can be as formal as a focus group or as economical as a feedback form on the service provider's Web site. Some Internet hosting providers compute visitor statistics to domain owners with valuable information such as most-visited pages, server hosting the visitor, the route the visitor takes through the site, and entry and exit pages. The prudent service provider will make allowances for the variability of less rigorous and less costly research. Using test markets is a worthwhile study technique to introduce new service offerings and especially unproven technologies. The scale of investments undertaken following broad primary research results should not be out of proportion to the thoroughness of the evaluation.

Focus groups bring together half a dozen to a dozen customers for a session that typically lasts a few hours. They are especially useful in determining a brand's position in the mind of the customer. Telephone or on-site surveyors often do not have the time or the skills to probe beyond the short answer. Focus groups can spend enough time on a single topic to glean qualitative dimensions of the brand position. Its disadvantage is that focus groups lack in breadth what they gain in depth, and running enough sessions to gain a representative sample of customers can be very expensive. Furthermore, the qualitative results that are the advantage of focus groups defy quantitative coding for statistical analysis. Moreover, developing market strategies based on the findings from focus group sessions is analogous to shaping political initiatives from focus group results. In both cases, the criticism is that leaders should not take direction from the public. In any case, focus groups are best applied to assessing corporate image and existing products, rather than evaluating new technologies. Key success factors include very clear session objectives and the selection of a skilled moderator. One-on-one interviews with customers offer more depth, but at more cost and narrower applicability.

Observational techniques plant observers in customer service situations, either to test the procedures through mystery shopping or simply to sample customers' reactions. Third-party observers often produce more accurate depictions of service transactions than do the customers themselves, especially because customers are often consulted long after the transaction took place. Observers focus on what service representatives do, not what they say that they do or what a training manual tells them to do. Mystery shoppers are trained professionals who interact with sales or customer service personnel with a fictitious problem or need, then report the quality of their service experience.

Customer intercepts involve stopping potential customers on the street, especially to demonstrate a new service offering and obtain feedback. While this technique is labor-intensive, the interviews can last only 5 minutes and provide quantitative data for later analysis. As a bonus, the interview educates the customer, possibly using some of the time to demonstrate the new feature's advantages. Thus, the customer intercept can act as a sales call.

The risks of failing in a new market are substantial, and an unsuccessful service launch can damage a service provider's brand equity at a cost well beyond the investment in the doomed service rollout. Unraveling a failed technology costs well more than the removal of network assets, as the highly competitive industry now demands considerable investments in advertising, promotions, and inventory in the case of wireless services and cable or broadband entertainment services. Test marketing enables the service provider to control the cost, the deployment and the potential damage of an untried service in an actual customer market. Selecting the market is critical. A data-mining venture through the service provider's customer database will reveal a community of price-insensitive innovators and early adopters who can help to uncover the weaknesses of the service and act as opinion leaders, enticing the early majority to emerge. If the test market is successful under ideal circumstances, then additional markets might have potential. Focus groups and customer surveys in the test markets will reveal issues that might not be a barrier for innovators, but which must be resolved before mass deployment takes place.

Concept testing is a related discipline. Marketing consultants can help service providers from early-stage concept evaluation by potential customers, shaping the new service to broaden its markets, evaluating purchase criteria utilized by the targeted market segment, developing predictive models that estimate the market for the service, in bands of prices, frequency of use, and demographic targets.

4.3 Yield Management

The discipline of yield management, also called revenue management, is simply the meticulous management of revenues in service industries where the merchandise is very perishable. The telecommunications services market, like the airline and hotel markets, needs to sell at capacity at every opportunity, or the revenue from an inactive circuit, an empty seat, or an unsold room, is gone forever. The hotel and airline industries have been optimizing their pricing with sophisticated yield management tools for two decades, while telecommunications service providers have chosen instead to fight price wars. Ironically, yield management was most likely first seen in the telecommunications service industry when it was under intense regulatory scrutiny.

The antiquated "value of service" philosophy that governed rates under monopoly pricing discriminated against certain target markets. Businesses paid well more than costs for their telecommunications network access, as did urban dwellers where dense facilities enjoyed economies of scale. Long-distance prices were several multiples higher than their costs, especially during the day when business customers had no choice but to pay inflated prices. In those days, price discrimination had a common benefit to all customers. To oversimplify, when businesses overpaid, consumers got affordable access. When consumers had phones, they could call businesses and buy from them. Until deregulation drove prices towards their true costs, business customers were content if they were exploited and earned customers in the process.

Price discrimination is disappearing as regulatory structures dissolve, as services like telecommunications (regulated), cable services (regulated differently), and Internet-based transport (not regulated) converge, and as competition presses willingness to pay closer to actual cost. But the airlines and hotels have proven that price discrimination in unregulated markets is not only possible but also desirable to optimize assets in intensely competitive services markets. Yield management is the mathematical means to achieving that optimization.

The notion of price discrimination in telecommunications services has undergone a metamorphosis. Only a protected monopolist can demand that its customers self-identify as business users so that they can pay more for the same service. In a competitive market, when one service provider practices unilateral price discrimination against a certain class of customers, all the customers in that segment will flee to competitors. MCI developed its critical mass by serving business customers and high-volume long-distance callers in the Chicago to

St. Louis market. Screening for yield management, based on customer behavior instead of customer type, is much more effective.

The service provider that profits from yield management has several characteristics. First, there is a fixed capacity and high fixed costs. Thus, the facilities-based service provider has more to gain from yield management than the reseller. But the reseller's suppliers are also facilities based. The reseller who manages its traffic to optimize its supplier's assets can benefit from lower costs of service. Fixed capacity will ensure that the last customer to demand network service will be satisfied, but the next customer after that will be blocked. Fixed costs set a standard for optimal profitability that will not be achieved unless the service provider sells every available minute or a target not far from full utilization.

An unbalanced demand for services is another indicator that yield management can be successful. The business segment continues to consume more minutes of service during the business day than any other segment. For the most part, business customers have no choice but to make calls during the hours that someone will be there to receive them. Thus, long-distance providers have tried to move network traffic originating in other market segments to off-hours and weekends. Most of this encouragement was in the form of discounts, though the desire for pricing simplicity and generally low long-distance prices have reduced these promotional structures. Instead, service providers targeting consumers—or targeting businesspeople during the hours that they too are consumers—have offered unlimited or nearly unlimited weekend usage as an enticement to gain subscribers. But the challenge for a particular service provider remains. After all, every service provider has the same network utilization problem: too much use during the week and not enough during the weekend. So a weekend promotion can divert traffic to a more convenient time, but it does not differentiate one service provider from another.

Indeed, the process of price discrimination through yield management actually benefits the customers in every market segment, to some extent even those who bear the brunt of higher discriminatory prices. The reason consumer telephony was so affordable a century ago was to ensure that a business with telephones could find its customers on the network. Thus the business benefited and so did the customer. Even today, the load of network traffic enables service providers to justify upgrades much sooner than they might have to serve only select, wealthy customers.

Yield management not only includes selling the most available services, but selling the services at the highest price the market will bear. After all, the concept of "yield" implies that the service provider is seeking the greatest possible disparity between expenses and revenues. Price wars in the telecommunications services industry indicate that competitors are very interested in volume and network utilization. Either they believe that the incremental minute, which is all profit when fixed costs are high, is worthwhile at any price, or they are uninformed or unwilling to use more sophisticated yield management techniques.

Yield management attempts to identify and separate those customers whose demand is price-insensitive from those who will buy only when price for a specific service is below a certain threshold, that is, to distinguish between inelastic and elastic demand. These extremities can be depicted on a continuum upon which customers at one end will pay anything for, as an example, wireless service, and customers at the other end that would not buy it at any price, as shown on Figure 4.1.

Somewhere between the extremes of customer willingness to pay is the line identifying the spot above which price discrimination can begin. Drawing the line too close to the high-paying customer gives up the extra revenue these customers would have been willing to pay. Drawing the line too close to the low-price seekers risks an abundance of unsold inventory.

At times, it is appropriate to sell at a deep discount to a customer who would not have otherwise made a purchase, when there is little or no risk that customers paying full price will learn about the discount, or when the high-paying customers would not be willing to accede to conditions the low payers will accept. In the airline and hotel industries, it is possible to tell a customer that a seat or a room is simply not available at a low price. Telecommunications service providers and Internet service providers cannot keep the customer from making on-demand calls. Instead, they can enter into agreements with their customers to balance the load on the network, and even agree to penalize the customer with very high rates for each transgression.

For example, the utilization-conscious facilities-based service provider could arrange highly discounted service for customers willing to make all calls off-hours, or make—and even receive—a very limited number of calls during business hours. The customer for such a deeply discounted but limited service, for example, is the ideal candidate for a so-called lifeline service. Like the otherwise irrelevant Saturday-stay requirement imposed by airlines and hotels to limit

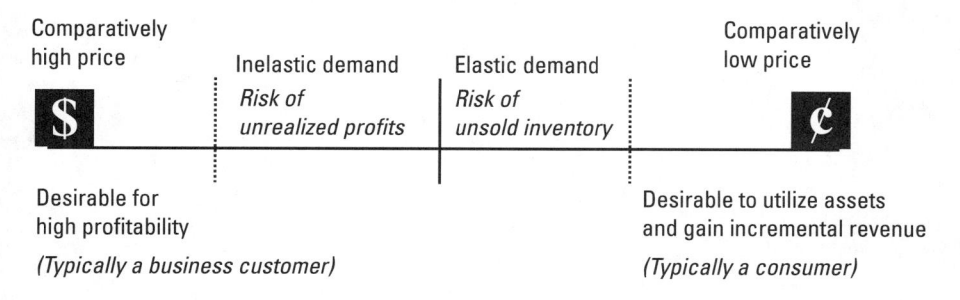

Comparatively high price	Inelastic demand	Elastic demand	Comparatively low price
$	*Risk of unrealized profits*	*Risk of unsold inventory*	¢
Desirable for high profitability			Desirable to utilize assets and gain incremental revenue
(Typically a business customer)			*(Typically a consumer)*

Figure 4.1 Yield continuum.

their deepest discounts to consumers, a lifeline service that limited discretionary calls to certain unattractive times would enable its subscribers to make emergency calls at payphone rates, keeping all but those in need away from the service. Creative regulations—or lack thereof—could leave it to service providers to offer lifeline, yet marginally profitable services in a completely deregulated market.

Reference

[1] Morgan, Martin, "Unearthing the Customer," *Telecommunications International Edition*, Vol. 37, No. 5, 2003, pp. 25–26.

5
Customer Demand

5.1 Capital-Intensive Industry Demand

The modern telecommunications service provider takes advantage of its marketing expertise to forecast customer demand for its services. Demand management considers both the creation of supply and its timing. Planning capacity for a demand for services needs to take into account both the capacity of facilities to meet customer needs and the lead time to get the capacity into service. Planning for the demand for products such as wireless devices needs to account for the number of devices and the lead time to replenish after reaching the restocking level. These efforts measure the "pull" of customer demand, but marketers are also responsible for the "push" of creating demand when networks are underutilized or services are in the final phases of the life cycle. In the past, push was the main planning function, to buildout capacity until networks could be profitable and manageable, and sell accordingly. Sophisticated forecasting techniques and customized service offerings provide an opportunity to respond to the pull of demand by creating supply to meet customer requirements as they arise.

Marketers participate in capital and expense planning on an advisory basis because they are well positioned to estimate short and long-term customer requirements. This process is as important as any other marketing function because it directly affects profitability and the long-term viability of the customer base: overestimating demand leads to stranded investment and unwelcome force reductions, underestimating demand leads to network congestion, lost revenue, and customer dissatisfaction.

Like so many other customer-centered planning and management functions, a focus on customer demand represents a turnabout from the supply-side planning that was the norm under a monopolistic industry structure. Monopoly service providers were able to control the timing and pace of technology introductions, managing the pricing of new and most often discretionary services—in conjunction with regulators—for optimal network utilization and socially acceptable profitability. In a competitive environment, customers enjoy indirect control over the pace of technology introductions and direct control over their adoption. In response to customer interest, competitor tolerance for narrow profit margins controls the prices.

Local telecommunications services competition began to emerge in the United States in 1996, assisted by the Telecommunications Act, and inspired by similar regulatory reform in Europe and around the world. At the same time, consumer spending on telecommunications was growing faster than any other category [1]. Many telecommunications professionals and industry analysts had every reason to be bullish about growth and demand. Mobile telephone growth, while sluggish in the United States compared to the rest of the world, was in double digits nearly everywhere. The Internet had begun to acquire a critical mass of users worldwide, and a tendency to forecast unprecedented growth was intoxicating. From 1997 until 2003, Internet traffic approximately doubled each year, an enviable growth rate on its own. Nonetheless, the rule of thumb that began to permeate industry conversations was that the doubling transpired every 100 days, a legend that was overly optimistic by more than three times the actual rate, and then the fictitious rate was compounded. The response of telecommunications service providers to such optimism was to repeat it without much skepticism, and then build networks to meet the demand implied in the projections. The result was to exacerbate rather than abate the negative effects of the industry slowdown of the late 1990s and early 2000s with unneeded investment in data transmission capacity. One estimate stated that data networks were operating only at about 10% to 15% capacity, as compared to the voice network at more than 30% [2], and that utilization of data network capacity in early 2003 might well be less than 10%. Furthermore, during the same period, network efficiency improved by a large margin, meaning that the initial investment in fiber might have already been perfectly capable of handling even the annual doubling of traffic.

The telecommunications services industry is capital-intensive, in that it maintains a large investment in fixed assets relative to sales, relative to employees, and relative to noncapital expense. Indeed, the continued advancements in software-based technologies inherent in the network, in customer care, and in back-office administration, increased the amount of investment in business capacity, not just network capacity. While computers and equipment will never replace all of labor in the investment and expense mix, it is possible that someday planning will assume that all costs are fixed, and that the capacity needed is simply a function of projected demand, timed to coincide with the availability of the network. Once committed to a given capacity, the marketing function will face the challenge of filling the space with revenue, exploiting economies of scale to reduce prices, exploiting economies of scope through convergence and vertical integration, exploiting economies of density by spreading the network, and simply increasing revenue through product development, pricing, promotion, packaging, and distribution.

Some debate exists surrounding whether the telecommunications industry genuinely supports sufficient economies of scale to give the largest facilities-based service providers significant competitive advantage. Scale economies enjoyed by monopolists might not be as great for competitors who are not guaranteed a clientele for their fixed-asset investments. Monopolists are more than willing to invest when rates are based on specific, risk-adjusted returns on their investment. Moreover, there is some evidence that economies of scale pass their maximum when the administrative burdens of managing an enormous enterprise exceed the benefits of business volume. The unbundling of the telecommunications services industry, through selective and phased deregulation, have also created business niches that are less capital-intensive than the industry as a whole and thus benefit less from economies of scale.

Economies of scope occur when the service provider integrates horizontally or vertically. Examples of horizontal integration include the provision of voice and data services through the same infrastructure, wireless and landline services, or offering narrowband and broadband alternatives to customers. Vertical integration such as enhanced services, content, and software-based network management capabilities, enable the service provider to reduce its revenue dependence on services dependent on high asset utilization. The service provider gains a competitive advantage from economies of scope when customers are seeking one-stop shopping, bundled services, or a single bill or customer service organization for an array of telecommunications services.

Economies of density arise when a service provider can reduce its costs, target existing customers efficiently, or reach prospective customers simply because of its established presence in the market. High-population areas generally support shorter access lines and better network utilization, creating lower prices, more demand, and lower prices yet. Economies of density favor the incumbent service provider over the new entrant, because the cost of the incremental customer is far less than the cost of the infrastructure averaged over a base of as-yet unsold customers.

In capital-intensive industries like telecommunications, estimating accurately the demand for services is essential, because the price of being wrong is very high. Predicting demand for telecommunications services is especially important for certain categories of investment. For example, predicting transmission demand is largely an aggregated function, that is, in this marketing function, there is no need to segment customers. Overestimating evening usage for airline reservation systems can offset unexpected growth in Internet usage by consumers, for example, and indeed, pricing structures attempt to equalize network traffic between customer segments. Other types of overestimation are more damaging. For example, the access line from the curb into a house is useless and stranded if a subscriber—or a substantial part of a segment of customers—decides to abandon fixed-line access in favor of wireless. Furthermore, in most places, the monthly rate subscribers pay for wireline access is still heavily regulated, even—or especially—in markets where resellers compete by marketing unbundled elements of the local loop. The formula for calculating the useful life of the asset and the capitalized labor assumes a long, long life for the subscription, resulting in low monthly rates and a slow recovery of the investment. Unlike transmission capacity between network hubs, which can be reassigned to handle unpredicted demand, finding an alternative use for stranded access investment is difficult. Both the price of fiber and the labor to install it are decreasing from historical amounts, especially on a bandwidth basis, but the anticipated future life of the investment will have much less to do with the longevity of the facility and much more to do with customer demand and competitive alternatives.

Increasing capital-intensiveness, the rise of fixed costs, and the overall automation of telecommunications networks might make a significant impact on the large business customer, a market segment that was historically important to the telecommunications service provider's profitability. Large customers are enticed by the notion of controlling, or at least managing, their costs through capital expenditures that bring expensive functions in-house. In the past, one major disincentive to building and maintaining an in-house network was the need to maintain a large technical staff with skills that did not match the core business of the enterprise. As networks become more automated, and labor costs become less onerous, service providers will find themselves in competition not only with other service providers but also with their own customers bringing the local portion of the telecommunications network in-house. Successful telecommunications service providers will offer proprietary technology for network management and superior customer care, enabling their customers to gain the ability to self-provision, maintain network visibility, and control network performance. In the end, all enterprises would rather invest in their core businesses than in support services, and the large business customer will trust its network to a service provider whose commitment to its own core business is unqualified. Similarly, opportunities will arise for telecommunications service providers that choose not to build facilities of their own to develop partnerships with enterprises who are willing to invest in the network equipment but reluctant to manage the network in-house. These unencumbered network management service providers can be equipment-independent and network-independent, characteristics that might be especially attractive to their potential customers.

5.2 Technology Industry Demand

The technology industry, including the telecommunications services sector, experiences pressure from customer demand beyond that in other industries. First, technology businesses are intensely competitive, so investing to meet anticipated demand for services not yet on the market—adding to the growth routinely occurring for existing technology-based services—is necessary simply to remain viable in the marketplace. Technology lines of business are so interdependent that services companies cannot readily integrate horizontally to related but countercyclical businesses. When one telecommunications sector—or even the computing technology sector—falters, the industry moves in lockstep. When the technology bubble burst, the staid telecommunications sector suffered alongside.

Moreover, stranded equipment and excess inventories are at greater risk for technology companies than in industries where assets remain at the same price and enjoy a longer life before they become obsolete. Overbuilding a fiber network that remains unused or below capacity results in the loss of immediate revenue needed to repay debt or investment. Compounding the problem, by the time the fiber or other asset is at a desired utilization level, the replacement cost is much lower, signifying a lost opportunity to delay its deployment and enabling latecomers to the market to enter at a lower price. Adding insult to injury, the cheaper replacement facility or end equipment is likely to have more functionality and lower installation costs in addition to its lower cost. Thus, while the services are perishable, even the timing of the infrastructure buildout demonstrates decay or at least obsolescence.

To their benefit, technology companies often support state-of-the-art information systems, consistent with their core business offerings. These information systems are vital to capturing and manipulating the needed data to develop accurate algorithms and demand estimates, most often based on historical information, such as adoption rates, customer price sensitivity, and external factors affecting demand, all resident in corporate databases. The telecommunications service provider needs to make the investment in integrated systems, demonstrate the discipline to collect the data, perform the required analysis, make realistic interpretations and forecasts, and revise the planning algorithms by comparing actual performance against analytical predictions.

Demand management realigns the energies of the enterprise to react to customer requirements. For a telecommunications service provider, approaching an infrastructure-on-demand philosophy is not as straightforward as it would be for a retailer, or even a manufacturer. Software to assist manufacturers and retailers in managing the supply chain, for example, has become quite sophisticated, and telecommunications service providers began to recognize the merits of automated supply-chain management (SCM). While much of the competitive advantage of automating the supply-chain management function is in cost reduction, SCM automation can offer marketing advantage to the enterprise as well. First, SCM can improve customer service by linking the availability of network services to customer demand and reducing the provisioning cycle. Second, supply-chain management systems provide a conduit between the enterprise and its suppliers, enabling suppliers to manage their logistics, inventories and other elements of the supply chain with the service provider, and with its business customers, for which the telecommunications service provider is a supplier.

5.3 Price Elasticity and the Demand Curve

For any product or service, a mathematical equation and a graphical curve can depict the relationship between its price and the quantity demanded. A service demonstrates price elasticity if demand falls when the price rises and demand for the service increases when the price drops. Demand for an inelastic product or service, by definition, does not change much if the price rises or falls. A rather complicated formula calculates the degree of elasticity from variables that are available from ordinary corporate recordkeeping. When the ratio determining elasticity is greater than one, it signifies that demand for the service is very receptive to price changes. As one example, for wholesale carriers, the price elasticity of bandwidth is between four and six; that is, a 50% reduction yields a 200% to 300% percent increase in revenue [3], and the price of a unit of long-distance bandwidth has dropped five hundred fold since 1996.

When elasticity is less than one, it means that a price change will not create a commensurate change in revenues. Services with low demand elasticity are not good candidates for price reductions or price wars because revenues will fall with volume, but raising prices will increase revenues, and, in the absence of higher costs, increase profits as well.

Unitary elasticity, in which a change in price has no net effect on revenues, means that the elasticity of demand is equal to one. Service providers need to monitor the location of unitary elasticity, because total revenue is maximized at this point. A service provider should lower or raise prices until the demand achieves unitary elasticity, and then stop until elasticity changes again.

Demand elasticity is receptive to other factors. For example, if substitutes are available for the service under study, elasticity rises. Thus, broadband service providers, including cable, local service DSL, and satellite-based, though they might each have a facilities monopoly in their own markets, create some elasticity of demand upon each other, though the services are not identical to customers.

The share of the customer's budget consumed by the service might also create elasticity. For example, a consumer of wireless services might be willing to concede to a 10% increase in the price of an SMS (short message service) message, but a business customer with a huge network would seek alternatives if the service provider made a surprising price increase to the price of a T-1 facility. Time is another factor: the customer who accepts a price increase in the short term grudgingly will find substitutes in the long term. Customers who view the service as essential will demonstrate less elasticity of demand; the demand of those who consider the service a luxury will be more elastic.

But demand elasticity has its limits. The service that is a complement to another purchase is affected by the demand characteristics of the related purchase. Growth in Wi-Fi connections in public places such as coffeehouses, after all target coffeehouses are wired, could depend more on the continued popularity of coffeehouses than upon the intrinsic demand for wireless connections. Cross-elasticity affects the demand for one service when the price changes for another service. Customers abandoning fixed-line service in their homes for wireless do so partly because the price of wireless services dropped significantly, an example of cross-elasticity. Cross-elasticity also reduces demand for one competitor's service when another competitor reduces its own price. Income elasticity occurs when a unitary change in the customer's income correlates with a change in demand. For example, satellite telephone services, in-flight telecommunications services, and some international calling demonstrate income elasticity. Differences in demand correlating with differences in income might provide worthwhile criteria for targeting certain services to a receptive customer segment.

Furthermore, even elastic demand becomes inelastic at the bottom of the demand curve. When bandwidth costs so little that customers use it like water—water being an excellent example of a commodity so inexpensive that few customers actively manage their consumption—lowering the price of bandwidth will not increase demand. A side effect of a very low price is that branding the service is less effective, and indeed, differences between two low prices mean much less to customers on a percentage basis than did equivalent differences at higher amounts. The customer who shops for a new long-distance provider to reduce the price from 30 cents a minute to 27 cents a minute is less eager to investigate alternatives between 3 cents a minute and 2.7 cents per minute. As all of these factors operate simultaneously, the technology-equipped service provider needs to perform the appropriate analysis routinely, taking as many factors as possible into account.

Finally, telecommunications demand for voice services, unlike the exuberance over data services, was historically underestimated [4]. Calculating price elasticity in a stable market, such as a highly regulated, slowly changing telecommunications market, was uncomplicated. Over time, price elasticity for ordinary telecommunications services approached homeostasis at 1, though newer data services supported a higher price elasticity of about 1.5. Even under monopoly regulation, the introduction of new services increased usage and caused a persistent underestimation of demand. Competition complicated the equation by increasing the pace of new service introductions. In this environment, users demand faster, more feature-rich applications, and software developers design for both present and anticipated hardware capabilities, desires that are met unfailingly by hardware manufacturers. Therefore, one way to compute elasticity of demand in the presence of unknown new applications is simply to assume that applications will drive up volume, technology advances will drive down prices, and price elasticity will take over from there.

5.4 Forecasting Demand

Demand forecasting includes conventional sales forecasting, but it goes well beyond it, and information-based tools can add to its sophistication and accuracy. Some elements of demand forecasting have been part of the process for decades. Seasonality, or even time of day or week, affects demand for telecommunications services, including local calling, long distance, and wireless. Though Mother's Day is the annual moment of highest holiday calling in the United States, the distinction of busiest calling day goes to the Monday following the long Thanksgiving weekend. Network capacity planning needs to take performance at peak usage into account. Telecommunications service providers have attempted to stabilize network utilization by inducements to price-sensitive customers to delay their nonessential calls until times when the network is underutilized. Nearly every wireless service provider offers abundant free minutes on weekends. Long-distance providers offer free minutes after business hours. Sprint offered "Fridays Free" to small-business customers, a program that was so successful at altering network utilization that Sprint abandoned it for lack of profitability. Clearly, Sprint had hoped for a complex elasticity profile—that the service provider would gain customers from its competitors at a higher rate than the losses it incurred from lack of Friday revenue. Instead, its existing and newly subscribed customers were happy to defer too much of their calling to Fridays, including displacing calls that Sprint had assumed were time-critical and revenue-producing at full price. If, though its high network quality, or customer care, or attention to the specific needs of small business, Sprint had kept those small-business customers after the promotion had run its course, then the promotion would have made sense. If not, then it was simply an example of poorly forecasting the demand characteristics of a targeted market segment.

Pricing structure can affect demand beyond a straightforward elasticity analysis. The service provider that offers contractual commitments at volume discounts will have larger but less frequent orders. Competitor prices can reduce demand temporarily, and if the service provider chooses not to match the competitor's price on a commodity service, demand might decrease permanently. Indeed, any substitute, whether offered by a competitor or as a new service offering, might affect demand. Transport on IP facilities is predicted to have a chilling effect on older transmission facilities, for example. New services that use existing network facilities more efficiently strands facilities formerly required unless growth takes up the slack. The pace and direction of regulatory activity affects demand as well, such as the introduction of number portability. Deadlines to comply with regulations affect the timing of demand, such as the requirement for wireless telephones to include E911 information. Any known factor, like these, can be included in a forecasting algorithm.

Many criteria affecting demand for specific telecommunications services or telecommunications services in general are outside the control of service providers individually or collectively, but some of these can be predicted and included in the statistical analysis. An economic downturn, such as the one experienced by many world economies at the millennium, will inhibit business and possibly consumer spending on discretionary technology or investment in general. On the other hand, business telecommunications demand will rise during weak economies in areas where telecommunications services serve as substitutes for other expenditures, especially air travel. Both videoconferencing and traditional calling serve as substitutes for travel in slumping economies. Call-center service bureaus might mitigate lower overall business volume because they serve as substitutes for in-house call centers in times of economic downturns—because companies resist making long-term investments—but they also thrive in times of prosperity, as call-center volumes rise with overall business volumes, for sales and for customer service. Enterprises bring the call-center function in-house only after it has reached a critical mass of volume. Population is another external indicator of potential telecommunications demand, as are housing starts for local service providers. The age of the population, information about which is available at no charge from government sources, can also provide broad direction to both market segmentation strategies and demand forecasting. One-time events create pockets of demand that are difficult to forecast but essential to serve. Natural emergencies, such as weather disasters, create a significant but temporary upsurge in demand, especially for wireless services, and present an opportunity for service providers to fail dramatically and publicly, and an opportunity to succeed heroically with some advance planning and provisions for disaster preparation.

Forecasting techniques can be quantitative or qualitative, and software is available to support the complicated statistical analyses they require. One technique in common use is regression analysis, which creates a formula to explain the sources of variation in an array of data points. For example, a simple linear regression analysis could develop a relationship between time-of-day and calling patterns. Multiple regression analysis applies when more than one variable predicts the outcome, such as calling patterns determined by both time-of-day and geography. Another form of regression analysis, time series analysis, projects future demand based on historical experience. Time series analysis might use a moving average of several previous periods of time, a smoothing technique or weighting to minimize outlying experiences, or a least-squares statistical technique to fit a trend line to the actual past data. A time series can include variables such as prices, market share, competitor promotions, and advertising campaigns to account for and isolate the changes to demand that occurred in connection with each of the variables.

Time displacement applies well to telecommunications markets, where adoption rates vary by service around the world. Time displacement enables the service provider to estimate demand using the actual demand profile from a different market at a previous time to predict upcoming demand as the new market develops. For example, European marketers can incorporate actual demand and growth characteristics from American adoption of the Internet, and American forecasters can learn from historical European adoption of wireless. Europeans and Americans can learn from Japanese adoption of broadband wireless technology. Furthermore, to keep constant the cultural factors that differentiate continents, marketers in the United States, for example, can incorporate the demand characteristics of U.S. markets further along in deregulation in demand forecasts for markets that are currently under deregulation, sometimes in the same region or state.

Trend lines enable the forecaster to test predictions of data from, for example, three or four planning periods back against the actual results from a later period, for example, one planning period back. This type of forecasting is called *ex post*, or after the fact, referring to the fact that all of the uncertainty of the forecast has been resolved by history. Testing predictive models against actual results can improve the model's ability to account for the constant rise in demand for telecommunications bandwidth. *Ex ante*, or before the fact, forecasts predict a future result based on past data, though the uncertainty has yet to be resolved. While demand models do poorly at predicting which individual applications will spur demand for services, forecasting demand in the aggregate can foster better customer service, network availability, and response time. Figure 5.1 depicts these concepts within their relevant timeframes.

Software enables the forecaster to perform iterations to improve the predictive model, but the model is only as accurate as the underlying input, and only as prescient as the limitations of quantitative analysis. Indeed, the more sophisticated the model, and the more iterative the results, the more that the process will gain credibility on account of the expenditure of effort alone, creating an impression of validity that might not be as robust as it appears. In a Booz Allen Hamilton survey evaluating the success of supply-chain management system implementations, 45% of chief information officers said that supply-chain

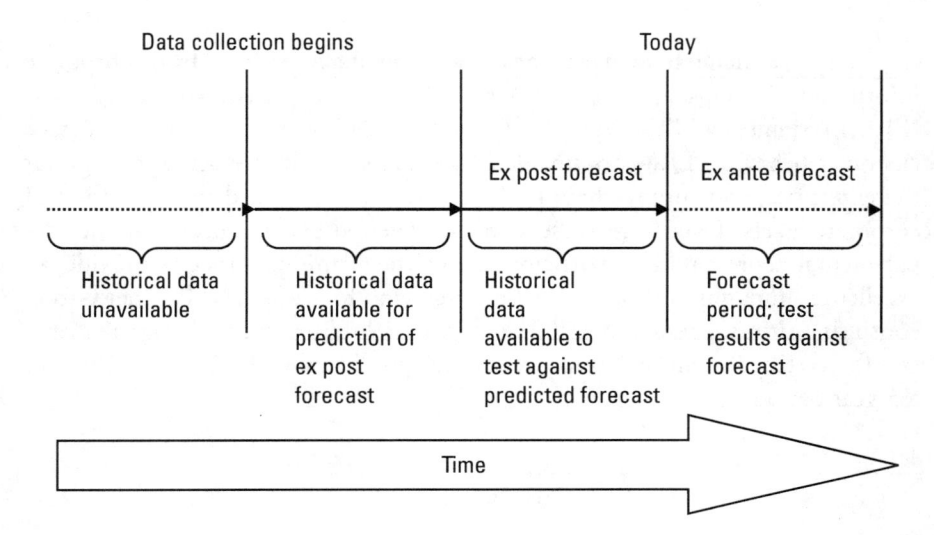

Figure 5.1 Improving demand forecasting.

technology had not met their expectations, and more than half placed the blame on demand forecasting software [5]. Moreover, while demand forecasting software makes feasible the inclusion of dozens of variables, the variation in the results can compound, creating large potential leeway for error.

One way to evaluate the strictly quantitative results is to add qualitative measures. One qualitative measure is simply to ask local experts in management for their reaction to the quantitative measures calculated in other analyses, soliciting suggestions that might improve the accuracy of the results. This method is called the Delphi technique after the Greek oracle, and forecasters would be wise to assume that this technique could produce results as flawed as its namesake priestess chanting incantations. But the Delphi technique is a way to assemble and structure the knowledge resident in the organization. Local managers who are close to customer activity can provide broad forecasts without using mathematical models, and other managers can then discuss and reconcile the differences that are bound to occur. Opinions from different corners of the enterprise, including network managers, sales managers, and customer service managers will help to rationalize the subjectivity inherent in forecasts. Some divisions are inclined to overestimate and others will underestimate, especially in relation to their own performance objectives. Qualitative measures have their place in demand forecasting, but they are best kept to a low profile or applied when quantitative techniques are weak: when no relevant historical data is available, when the time horizon is too long to make accurate forecasts, or when the market is so unstable that predicting even the past from actual results finds little guidance.

Finally, the best demand forecasts bring together data from corporate information systems, such as CRM records, enterprise resource management (ERM) systems, and SCM systems. Customer relationship management details customer behavior, ERM systems describe the flow of information throughout the enterprise, and supply-chain management systems record the utilization of corporate assets. Data from these sources, reviewed collaboratively by interdepartmental representatives participating in the planning process, provide an excellent foundation for robust customer-demand forecasts. The Gartner Group consulting firm suggested in a 2001 report that enterprises that integrate disparate forecasting systems will improve revenue predictability by 10% to 25% over a 3-year period.

References

[1] Standage, Tom, "Beyond the Bubble," *The Economist*, Vol. 356, No. 8345, 2003.

[2] Odlyzko, Andrew M., "Data Networks Are Lightly Utilized, and Will Stay That Way," *Review of Network Economics*, Vol. 2, No. 3, 2003, pp. 210–237.

[3] Gilder, George, *Telecosm: The World After Bandwidth Abundance*, New York: Free Press, 2002, p. 268.

[4] Lanning, Steven G., Shawn R. O'Donnell, and W. Russell Neuman, "A Taxonomy of Communications Demand," *Telecommunications Policy Research Conference*, Alexandria, VA, September 1999.

[5] Worthen, Ben, "Future Results Not Guaranteed," *CIO Magazine*, Vol. 16, No. 19, 2003, pp. 46–51.

6

Competitive Intelligence

6.1 Developing Competitive Intelligence

Service providers can apply data-mining techniques to internal databases to gain a detailed knowledge of their own services and their own customers' behavior. Nonetheless, a full understanding of the marketplace requires knowing about competitors as well as the industry trends that drive the market. This information is available through a formal or informal competitive intelligence program. In a confusing semantic twist, the phrase *competitive intelligence* refers both to the program of assembling data and to the knowledge that is the end product of the program. In a March 2002 survey, consulting firm PricewaterhouseCoopers found that 42% of fast-growth company chief executive officers viewed competitor information as very or critically important to profit growth, especially price changes, product initiatives, and changes to corporate strategy.

Most competitive intelligence is routine and persistent, and does not resemble the residual image left by novels and movies, of cloak-and-dagger meetings to trade corporate secrets, parking lot surveillance to calculate payroll costs and business volume, or rummaging through garbage cans to find documentation that crept away from the shredder. Indeed, the most dramatic forms of intelligence gathering mythologized in the media are actually espionage and are illegal. A service provider can keep busy and stay well informed without resorting to the risks of espionage. Virtually all of the information a service provider needs is in the public domain.

Strategic intelligence considers the broad future direction of competitors and is a particularly valuable form of competitive intelligence. Because of this, strategic details are most often shielded tenaciously from the view of the public. Nonetheless, broad direction and hints of details are available to the determined observer. The best sources of strategic intelligence are trade journals, especially interviews with chief executive officers and other senior management, and press releases, often available on a company Web site. Furthermore, the executive or market analyst that simply mulls over the abundance of reading material available for most major competitors will be able to surmise most of the competitor's defined strategy or perhaps an even better strategy that the competitor hasn't discovered yet.

Tactical intelligence is less predictive and more operational. Press releases on corporate Web sites are good sources of imminent activities. Trade journals report industry initiatives, such as regulatory developments, and trade shows can provide information in time for the customer-focused service provider to react competitively. Tactical intelligence is useful for formulating competitive responses to actions such as price reductions or promotions. Extremely timely information might enable the service provider to act preemptively rather than reactively.

Counterintelligence is the process by which the service provider attempts to protect one's own corporate data better than its competitors do. On the service provider's Web site, for example, a small registration form might dissuade anonymous information-seekers from delving further, while not presenting too large a barrier to legitimate prospects interested in buying. Furthermore, counterintelligence needs to ensure that not only corporate strategies, but also even the competitive intelligence developed within one's organization remains as a protected trade secret. When a competitor learns the extent of a service provider's competitive knowledge, the investment the service provider made in gaining that knowledge essentially becomes lost. The competitor revises its own market plans, and the service provider is back where it began, without competitive intelligence of any value.

The purpose of a formal competitive intelligence process is to gather knowledge and initiate appropriate actions concerning a defined set of enterprises that compete for the same position in the customer's mind and the same portion of the customer's budget. This process contains a series of steps, as shown on Figure 6.1. The first is the normal and systematic observation of industry and other events and trends that any enterprise should conduct routinely. This step will uncover areas for further analysis, which are documented and ranked in order of priority in the second step. The third step is the legal collection of information about the most significant competitive developments.

The competitive intelligence program adds intelligence to the data collection effort in the fourth step. In fact, the analysis portion of the intelligence program is the more important component. It evaluates the quality of the information gleaned, strives to anticipate what competitors will do before they do it and categorizes disparate information into usable taxonomies for management. The fifth step makes actionable recommendations. Management's responsibility

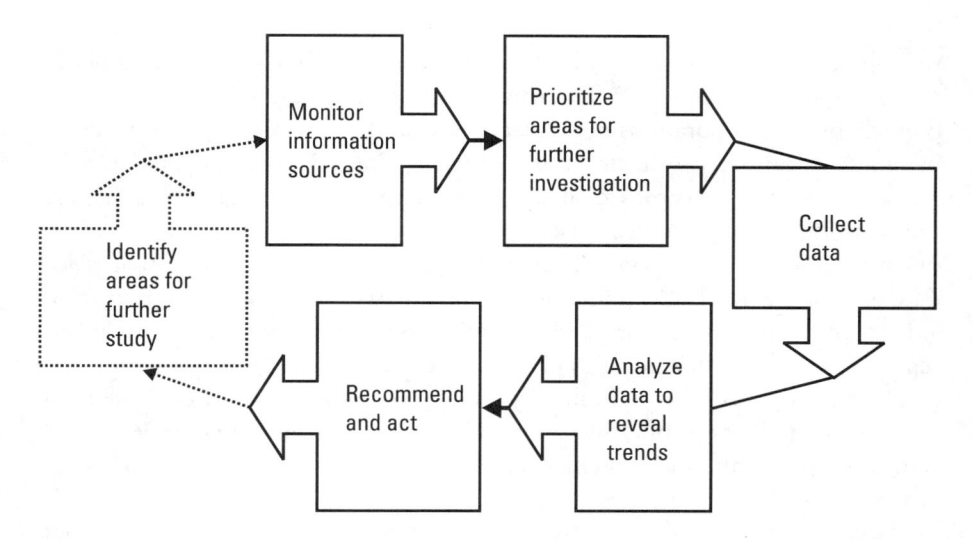

Figure 6.1 The competitive intelligence process.

is to plan and take steps accordingly and quickly in response to competitive actions or proactively to protect competitive positioning in anticipation of competitor actions. Without timely adjustments of strategy and tactics, a competitive intelligence program is not worth the effort. Simply adding the intelligence and action to competitive intelligence can place a small service provider at a competitive advantage against the huge corporations that subscribe to foot-thick, six-figure reports, only to shelve them in the corporate library after a glance. The last step reevaluates the intelligence and the results of initiatives to repeat the process against the changed competitive battlefield.

Some of the most valuable competitive intelligence resides inside the service provider's own organization, in the experiences of sales representatives, customer service personnel, technical support personnel who attend industry conferences, and management. One of the challenges of a competitive intelligence program is to develop a methodology for sorting the collected information in a way that it can be analyzed and retrieved when needed. Software support is available to assist in the cataloguing and keyword retrieval of competitive intelligence data.

Because of the Internet, it has never been easier to learn about competing services and their pricing. All competitors worth studying support up-to-date Web sites that advertise service packages in great detail, sometimes including their own planning data. For example, many U.S. local access providers support scripts that prospects can launch to learn when broadband data access or other new services will be available, even at a level of detail to include the wire center. Useful information about a company's direction is written between the lines of the job postings on their site or on help-wanted ads on-line in newspapers local to the corporate headquarters. Extensive financial information is available about publicly owned corporations in annual reports and financial filings on the company site or on the Securities and Exchange Commission's Edgar database. Industry analyst reports provide detailed pro forma financial statements that are in line with internal statements used by corporate management. A simple search of a newsgroup can lead to employee postings, which might give at least anecdotal data at a low level of certainty about morale or service plans. A service provider with enough persistence and ingenuity could almost generate a multiyear replica of a competitor's business plan. Of course, the market analysts employed by one's competitors are undoubtedly able to mine the same planning data on one's own site if they apply the same diligence, in a miniature corporate cold war. Indeed, competitive intelligence programs can be necessary simply to achieve competitive parity.

All of the knowledge is extremely time sensitive, which is the reason that the competitive intelligence program should gather data regularly, even in small amounts, rather than in an annual exercise. A corporate intranet or even an e-mail client can use easy-to-complete forms to encourage employees to provide competitive information as they find it. Outside contractors or public relations firms can help with the data collection by providing clipping services. Internet search engines can scan the news every day for developments in the marketplace generally or about designated competitors. Trade publications offer free industry news updates daily or weekly, delivered via e-mail at no charge. For a moderate charge, certain research and database sites scan journals and newspapers for updates in targeted industries or against selected keywords and alert their customers to news items of interest.

The customer-centered service provider will pay particular attention to the customer's view of competitors, whether the competitor is the customer's present service provider or if the competitor is simply a candidate for a future relationship. Thus, the normal venues for competitor information—on-line databases, government reports, benchmarking studies, trade journals, and the like—are necessary but not sufficient, because they report facts instead of the customer's opinion. Most of the major service providers have inspired newsgroups on-line, which any search engine can navigate. General newsgroups about various technologies or services also include posts about specific vendors. The data is anecdotal, and the people that bother to post their opinions on-line are likely to be seeking support or are otherwise unhappy with their present relationship. Therefore, the procedure is anything but statistically sound and probably demonstrates a negative bias. On the other hand, the newsgroups have thus already segmented the universe into less-satisfied customers, and the problems they voice provide potential differentiation strategies for an unmeasured market segment. One worthwhile by-product of learning about competitors from their customers is that the analysis will uncover unmet needs that lead to new business opportunities. Studying competitors alone will only produce ideas that are already in the marketplace. Listening to customers presents an opportunity to innovate.

Setting goals for intelligence gathering is vital, because goals that are too vague will not be useful, and goals that are too precise will not bear fruit. An overly broad goal of profiling a handful of known competitors does not require a professional to conduct the analysis; a one-page description of the service provider on any Web portal will suffice as a general description for management. Similarly, an excessively specific goal such as defining what the next service launch will be for the market leader or exactly when it will occur would take too much effort or border on espionage. As occasionally happens during the unstructured data mining of customer records, some of the most valuable finds in competitive intelligence analysis emerge from a general review of data, not as the result of a specific inquiry. The ideal pace is to monitor competitor Web sites, news sources, and trade journals and spend the majority of effort on analysis of the news items that are readily available.

Trade shows where service providers are the exhibitors, not the attendees, provide a saturated landscape for competitive intelligence gathering. Because of the potential for news coverage, service providers save their largest announcements for major shows. Half of the attendees are seeking new solutions and suppliers, and one-third of the participants are presidents, owners or executives with decision-making power [1]. These characteristics already provide good reasons to set up a booth and sell at a trade show, but it also presents an opportunity to send some additional employees who are dedicated to being competitive intelligence gatherers. While the trade show sales force is selling and training, these individuals can conduct surveys, arrange face-to-face meetings to get direct feedback about customer needs and their image of competitive services, and interview reporters from the trade press about their views of the competitive landscape.

Role-playing one's competitors is one way to find creative avenues to monitor. A service provider that is comfortable in its relative dominance over a competitor's market share needs to realize that the competitor following behind is quite uncomfortable in the present position. The market leader should attune its competitive intelligence program to potential merger activities that could reshape the marketplace. Similarly, rumored merger negotiations between a willing party and a less-willing one might signal an opportunity for a white-knight proposal. For example, in the late 1990s, WorldCom and Sprint were the number two and three long-distance providers, with 26% and 11% market share, respectively, to AT&T's 43%. No other long-distance provider was close, with the largest market share of the remaining competitors at 2% [2]. Sprint had no hope of catching up to either of the two leaders, and would need to pursue a flanking or guerilla strategy to be successful. These strategic alternatives would be unappealing to Sprint, an incumbent service provider that was accustomed to enjoying a monopoly in its franchised territories and a large investment in a national fiber network. WorldCom had little hope of capturing much more market share from AT&T. Furthermore, much of the profitability and growth of its rivals was occurring in their wireless divisions, which also provided leverage for their long-distance transmission facilities. Thus, Sprint was without a significant market share, and WorldCom needed a wireless network. AT&T would have been irresponsible if it did not review its market positioning within a hypothetical universe against a single competitor with an all-fiber transmission network and a large wireless customer base. Their announced merger surprised many in the industry who had been mesmerized by the growing relationship between WorldCom and British Telecom. The merger between Sprint and WorldCom never took place, thwarted by regulators, but a shrewd practitioner of competitive intelligence never expects outsiders to rescue its corporate strategy.

One temptation is to spend too much time gathering data at the expense of its analysis. A competitive intelligence analyst will never achieve the comfort level that enough data has been unearthed, but diminishing returns after data becomes repetitive are simply a by-product of competitive analysis. The biggest challenge to the analyst is to make sense of the disparate forms and qualitative nature of the quantity of intelligence gathered. For management to derive benefit from it, reports need sorting into categories, discrepancies between facts need resolution, and gaps between data points need extrapolation. In the end, the two results that matter most are the quality of the conclusions drawn, however tentative they might be, and the soundness of the actions management takes on their behalf.

6.2 Finding Strategic Competitors Before the Customer Does

As if it were not challenging enough to monitor and react to the expected behavior of known rivals, an important element of competitive intelligence is to uncover and follow the marketplace movements of strategic competitors as well. A strategic competitor is an enterprise that is likely to enter markets of interest to the service provider, whether or not the service provider conducting the intelligence program is presently serving the market under study. In some cases, the service provider is monitoring the activities of another enterprise in a market that neither of the participants in the study is currently serving. For example, as late as 2003, though few local service providers were offering VoIP services regularly to their customers, they were all exceedingly interested in knowing who was committed to that market (generally small players) and which potential competitors to expect to enter in the future (generally large players). By 2004, announcements of VoIP investments by major service providers were common.

Often the entry strategy into a nearby market serves as leverage to the existing pool of customers. Strategic competitors might enter the market through vertical integration, such as America Online (AOL) entering into consumer broadband services through its Time Warner cable division. AOL, under pressure from other Internet service providers, found broadband access an opportunity to differentiate its service from its actual competitors who were not affiliated with broadband service providers. AOL did not need to be a spectacular performer in broadband data services, as long as it paid off by keeping customers and adding to profitability in its core business of Internet access. Apparently it did not perform even to that standard, as AOL gave up its broadband access business in 2004 to focus on its core content and services portfolio instead.

Strategic competitors might choose to enter through horizontal integration. Public utilities, who often have rights-of-way through large territories, constructed fiber networks in their territories to serve wholesale markets, and in some cases entered the consumer retail market with broadband service offerings. Williams Communications began as a gas and oil pipeline company, and leveraged part of its 100,000-mile infrastructure to build an 11,000-mile fiber-optic network in the mid-1980s using decommissioned pipelines. After selling that legacy network to WorldCom, Williams built a new fiber network to meet its future needs as a carrier's carrier. Strategic competitors can enter the market through an expansion of distribution channels. Both Qwest and Global Crossing originally professed to be carrier's carriers, and both changed their corporate direction to become competitors in retail markets. Even carrier's carrier Williams retained some large retail customers. Satellite television provider DirecTV entered the high-speed Internet access market to consumers through its DirecPC and Direcway brands. Last, competitive entry can come from geographical expansion. Germany's Deutsche Telekom, Vodafone, and NTT DoCoMo investment in wireless service providers around the world should convince domestic service providers everywhere to widen their sights territorially.

How does the service provider uncover these potential competitors? By definition, strategic competitors are not within the purview of those who follow a particular market. The first task is for the service provider to distance itself from its own preconceptions and its own market and look beyond it. The second is to identify those business opportunities where service portfolio gaps exist and where special skills or assets held by another entity would facilitate market entry. Last, the service provider should identify what specific enterprise or type of enterprise would succeed in the marketplace, or would leverage its existing businesses so much that only a modest success would be sufficient to improve its profitability. Wireless providers without intercity networks, for example, might have anticipated AT&T eliminating the distinction between wireline and wireless minutes when it launched its Digital One Rate campaign. Because most of the remaining wireless service providers were regional Bell operating companies, AT&T's competitors in most markets were not only unprepared to compete with the new pricing structure; they were prohibited by law from retaliating by building networks.

Distancing oneself from the market involves some investment and some organizational flexibility. Outsiders are useful because they might be willing to ignore conventional industry wisdom, or they might simply not be as immersed in it as managers who have devoted careers to a particular marketplace. Business students on internships, for example, can assist in the analysis. The challenge is to find someone who is close enough to the business to understand the issues, but not so close as to be blinded by knowledge and preconceived notions. Another source of entrepreneurial talent resides within the organization. Verizon sponsored an incubator program, in which managers with ideas for innovative telecommunications businesses could produce business plans and prototypes under the auspices of the corporate development division. A service provider whose culture tolerates dissent will be most receptive to employee insights about possible strategic competitors. Without creating a second corporate structure, service providers can benefit from duplicating some functions to create new services or innovate in selected critical business processes. At best, an initiative through which a service provider competes with itself can subvert similar efforts constantly taking place beyond the enterprise. At least, a set of best practices—or, at least, better practices—will emerge from one team or the other.

Whether a new competitor's market entry would succeed is only half the problem. Sometimes the simple act of introducing change shifts the whole competitive landscape. AOL was not the first Internet service provider, but its success rests in part of the pioneering efforts of earlier and much less successful ISPs such as CompuServe (now owned by AOL) and Prodigy (now owned by SBC). These early efforts created a small audience of innovators, tested a variety of pricing packages and browsers, and paved the way for advertising models for both ISPs and portals. Both of these also-ran ISPs came from the horizontal integration efforts in other industries: CompuServe was the invention of tax preparer H&R Block, and Prodigy was the investment of consumer products retailer Sears, broadcaster CBS, and computer manufacturer IBM. It is worth noting that of these pioneers, only IBM has remained in the ISP business, and that the ISPs they created have found their roots with infrastructure-based service providers.

Simply imagining the market entry of a service provider with unusual skills should prompt the service provider to look objectively at its own strengths and its weaknesses relative to enterprises in related or complementary industries. Moreover, the service provider needs to take special note of an outsider's assets that facilitate market entry and discourage competitive response. For example, the FCC classified early VoIP incarnations as information services, and thus not subject to access charges. Until a final ruling about whether access charges should be imposed on VoIP traffic or removed from conventional access lines, a dynamic situation like that presents an opportunity for a new entrant to provide local service at a competitive advantage.

Furthermore, the service provider can adopt a posture of preemption. Rather than wait for strategic competitors to invade the market, entrepreneurial service providers can seek out related markets, where their own skills and assets pave the way for success, leverage, and increased profits. At the very least, the strategic entry into nearby markets might distract the competitor's management with retaliatory strategies sufficiently to delay their own expansion into telecommunications services.

6.3 Perceptual Mapping

Knowing the positioning of one's own brand in the mind of the customer implies knowing one's position among a sea of competitors. One way to quantify the otherwise inexact measurement of competitive positions is to create a perceptual map. Based on opinion survey results, perceptual mapping documents the relative market positioning of an array of brands—or almost anything that can occupy a share of the customer's mind—along a series of selected attributes. A perceptual map is a two-dimensional chart that can show up to a dozen or more brand attributes at a time, in a format that is equally suited for a seminar for statisticians or the cover page for a board of directors meeting.

Perceptual maps depict the relative proximity of objects, whether measurable by physical distance or not. To picture a perceptual map, it is worthwhile to begin the discussion with a geographical map. For example, from a table showing the distances between 10 city-pairs in the United States, but not their direction from each other, a mathematical iteration could eventually create a pictorial map with the cities located in their proper physical positions relative to each other, although without the land borders. A similar positioning of brands, using "distances" such as percentage of respondents answering "yes" or "no" to a survey question or numbers along a continuum, such as low price to high price, will yield a perceptual map. Though manual computation is possible, it is not necessary, as software support enables the service provider to sort through as many variables as desired. The software can also present the results on an easy-to-interpret chart and perform sensitivity analysis. Further analysis might divide the survey data against market segmentation parameters, to highlight target markets, and identify competitive strengths and weaknesses in each of those markets.

The benefits of a visual competitive analysis are numerous. Locating one's own perceptual location within a group of desired attributes enables the service provider to target customers who think that the service provider excels at the very attributes that the customers believe to be important. The analysis charts the enigmatic domain of the mind by isolating the customers who are receptive to the service provider, whether or not the service provider really does excel in the attributes the customer values. The only measurements that matter are what the customer wants and the customer's image of the service provider, not what the service provider wants to be, nor whether the service provider is factually meeting the criteria inside the customer's mind. Companies peering at a cluster of competitors surrounding them might not be pleased to find that they hover inside a group of low-quality providers, or the low-value prices, but the charts simply report what customers have said that they think.

The proximity of competitors on the chart will depict the most looming challengers for the customer's allegiance, and the results might be surprising. A service provider that is third or fourth in its market might find that its charted position is diametrically opposed to the market leader, but that its market positioning is in head-to-head combat with another service provider that was previously unnoticed. In that case, a possible winning strategy might redirect marketing investments directed at the industry leader and point them instead at differentiation from the neighboring threat. The service provider perceptive enough to include strategic competitors in the analysis as well gains insight into their imminence, their threat, and potential differentiation tactics when outright competition becomes inescapable. The perceptual map also helps the service provider identify which attributes are best candidates for industry leadership and which would present huge obstacles to overcome. Suppose that a service provider perceives itself to be the low-cost provider, but its customer base sees it as strong on customer service. Furthermore, it might find that its position along the cost continuum is clustered with four other companies. This service provider might consider changing its targeting strategy to focus on its differentiating strength and abandon its fight for cost leadership.

The service provider that finds itself at odds with its own marketplace positioning can take a variety of actions to improve its location on the perceptual map. The first option is to reposition into a different spot through traditional marketing tactics, such as advertising, new service introductions, improvements to service quality, and any other relevant marketplace approaches. The distance between one's own position and those of its rivals is relative, so any opportunity to reduce the strength of the competitor's position is desirable. One method of accomplishing this is to alter the scorekeeping mechanism that the customer uses, either by reducing the importance of unfavorable attributes in the customer's mind or by diverting customer attention to attributes in which the service provider excels without competition.

Figure 6.2 compares the merits of advertising on various Web portals. Unlike ordinary mathematical charts, the *x*- and *y*-axes represent attributes, not amounts, and provide only a canvas to demonstrate proximity and distance between points. Those attributes that extend far from the origin are more important to customers than those that stay close by, no matter which direction they lead. Those service providers that hover close to an attribute are viewed by customers as having more of that attribute than their more distant competitors. Without stating it explicitly, customers demonstrated that they viewed attributes on opposite sides of the map as tradeoffs, such as "delivering value" (4) and "low price" (8).

The chart, from a 1999 survey, demonstrates that AltaVista and Excite occupied similar territory in customer perception, that customers believed them both to be results-oriented and delivering value to them. The finding is intuitive, as one might expect that AltaVista would have found Excite to be its most contentious rival in 1999, targeting the same customers. The plot shows that each of those brands would be the likely destination of customers who switch from the other one. Similarly, the attributes of "quality customer services" and "targeted audiences" were viewed as highly related, as judged by the small angle between the lines, and AOL appeared to have little competition in meeting these needs in its advertisers. Its performance against those attributes went well beyond differentiation, and its only competitor in the space, MSN, performed less well. The longest lines appear for those attributes that are best at differentiating the brands.

Note that the perceptual map is only a snapshot in time. The best evidence of that fact is that this chart does not include Google, a portal so ubiquitous only a few years later that its name became a synonym for search, a portal that

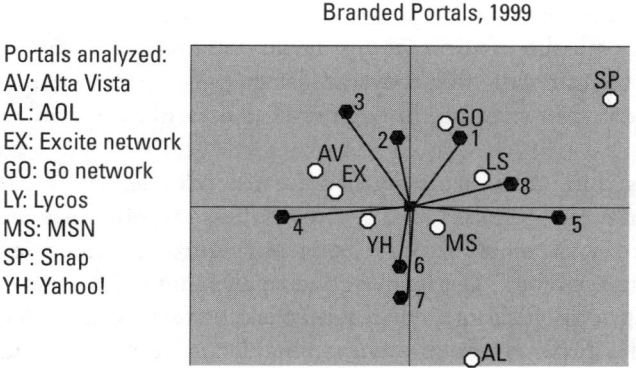

Branded Portals, 1999

Portals analyzed:
AV: Alta Vista
AL: AOL
EX: Excite network
GO: Go network
LY: Lycos
MS: MSN
SP: Snap
YH: Yahoo!

Attributes:
1. Measurement accuracy
2. Care about my bottom line
3. Result oriented
4. Delivering value
5. Purchase flexibility
6. Quality customer services
7. Targeted audiences
8. Low price

Most respondents were Web marketers or advertisers.

Figure 6.2 Example of perceptual mapping. (*From:* [3]. © 1998–2000 CMO Consulting International. Reprinted with permission.)

emerged from unknown to uncontested leadership in the space of only a few months. Indeed, creating a perceptual map at intervals is most useful because it demonstrates the shifting of market position before other metrics such as revenue loss or annual user satisfaction surveys reveal them. Sprint, for example, held the leadership position in "network quality" in both its long-distance and wireless markets for years before Verizon captured the attribute. Had Sprint been formulating perceptual maps routinely, early erosion of its leadership might have been apparent, and Sprint might have invested or marketed more intensely to retain its position, or conversely decided that ownership of the attribute did not translate to profitability. In that case, Sprint might then reallocate the resources positioning the brand as high quality into lowering its costs or improving customer service.

6.4 Competitive Parity

Competitive intelligence is worth its investment only if it leads to some action based on its findings. Service providers need to decide how they will react to, or act in advance of, competitor initiatives. Competitive parity refers to the process by which the service provider meets or dismisses marketing and other competitive initiatives by its competitors. While much of competitive response occurs after the competitor has already acted, competitive intelligence can offer several advantages when the service provider has advance knowledge of competitor direction, forthcoming changes to the marketplace, or customer desires. Once the most threatening competitors are identified, through ordinary intelligence techniques or by studying a perceptual map of the marketplace, a role-playing exercise can assist management in predicting the appropriate competitive moves.

The service provider who takes a risk by innovating can be rewarded with first-mover advantage. Whether first-mover advantage is absolute is a matter of some debate. On the plus side, the first mover might gain preemptive access to resources such as distribution channels or investment, or acquire proprietary technology that offers competitive edge, or retain its base of early-adopting customers by creating switching costs, or simply gain the network effect of a critical mass of customers that creates a viable market. (Network effects, also called network externalities, are examined in Chapter 15.) Nextel's introduction of push-to-talk service offered an example of first mover advantage and enjoyed several of its benefits. Its proprietary technology harkened back to Nextel's roots as the specialized mobile radio dispatch service provider FleetCall. Nextel's proprietary technology clustered customers in calling circles of opinion leaders, creating switching costs for customers once other wireless service providers offered similar features. For a customer to switch to another vendor, the customer's entire calling community would need to switch to retain the push-to-talk feature.

Thus, the first mover position created a community of customers and built barriers to new entrants; anyone wishing to gain a single customer who valued the push-to-talk feature would need to win over the entire related customer group. Network effects occur when the value of the service increases with the size of the customer base, clearly an advantage for high-growth first mover Nextel. The benefit of critical mass was compounded by the lag time before Nextel's competitors, including market leader Verizon, offered their own versions of the feature and other capabilities that reduced the power of Nextel's positioning. On the downside, there is some evidence of a second mover advantage. For example, later entrants to the market gain the advantage of an educated early-adopting customer who was funded by the first mover, and gains the later-adopting customer once the market enjoys reduced uncertainty and risk [4].

Nextel's push-to-talk wireless capability set the standard for other service providers. By the time Verizon launched its own version of push-to-talk, expectations for service quality were high. Industry observers criticized Verizon's call setup and latency, especially as compared to Nextel's service [5]. Nextel, in learning of competitive inroads into its primary service differentiator, had planted the seeds of failure for its rivals by asserting that other wireless providers would not meet the Nextel's standard [6]. Verizon's launch of push-to-talk technology probably inspired Sprint to accelerate its rollout of its own version of the service. But first-mover advantage would not have helped Nextel if its competitors had been able to offer lower latency—a criterion that Nextel itself had set—or if they had been able to offer the same network quality at a lower cost.

References

[1] Barron, Anne, "Three Easy Steps for Gathering Intelligence at Trade Shows," *Competitive Intelligence Magazine*, Vol. 3, No. 4, 2000.

[2] "MCI, Sprint Could Pass Antitrust Test," *Wall Street Journal*, Eastern Edition, Volume CCXXXIV, No. 61, 1999, p. B6.

[3] CMO Consulting International, http://www.webcmo.com.

[4] Lieberman, Marvin B., and David B. Montgomery, "First-mover (Dis)advantages: Retrospective and Link with the Resource-Based View," *Strategic Management Journal*, Vol. 19, No. 12, 1998, p. 1111.

[5] O'Shea, Dan, "Push-to-Talk Battle Erupts in War of Words," *Telephony*, Vol. 243, No. 16, 2003.

[6] O'Shea, Dan, "Verizon Confirms Push-to-Talk Launch," *Telephony*, Vol. 243, No. 14, 2003.

7

Channel Management

7.1 Traditional Distribution Channels

Distribution represents the steps needed to get services to customers, including transactions, logistics, and fulfillment. Distribution involves finding and motivating participants in the channel leading from production to delivery, covering geographical and vertical markets, and putting the service within reach of customers. Direct distribution refers to delivery from the producer or service provider to the customer without any intermediaries. Direct distribution is most common and very efficient under a monopoly market structure, and virtually all of the intermediaries presently in the service provider market entered the industry in the last 25 years.

Services distribution differs from product distribution in that the traditional roles of a wholesaler—to break bulk, to store inventory, and to resell the goods to retailers—are not part of the service delivery process. Monopolists controlled the entire distribution cycle for their services in the interest of the complexity of social pricing. As regulations slipped away, opportunities to sell using indirect channels grew. Because indirect sales channels did not require the enormous investment in network facilities that direct service provision necessitated, the distribution channel was an attractive point of entry into the market.

For previously regulated services, the indirect channel was pulled by entrepreneurs rather than pushed by service providers, who generally viewed resellers as competitors rather than partners and made little effort to cooperate with them. Wireless services, which came into their own in a hybrid market dominated by incumbent fixed-line providers and entrepreneurs who were less determined to own the entire distribution channel, used a mix of employees and agencies to sell their services. In the Internet access market, as a rule, Internet service providers distributed their startup software in electronics and office supply stores, and occasionally used their own direct sales forces to sign business accounts.

Within the category of indirect sales are brokers, agents, and retailers. Brokers facilitate transactions between parties, never really taking responsibility, and generally stay involved in the transaction only on a temporary basis. Agents represent a service provider either exclusively or within a large portfolio and might take on value-added functions such as account management, customer support, and system integration. Retailers offer a variety of products from numerous suppliers, but they provide the service provider with a personal relationship with the customer, and they perform vital marketing functions such as merchandising and promotion. Retailers can also offer services that will close the sale, such as credit or a strong brand image.

Many factors have transformed the traditional distribution mechanisms, many owing their roots to advances in technology. First, customers are more willing than ever to choose low-interaction indirect channels such as inbound call centers and the Internet, especially for products with which the customer is already familiar. Similarly, services have become more standardized, such as the tiered service offerings of wireless service providers at the same time the channel itself is increasing in customer value. Competitive pressures have reduced the technology life cycle, creating turnover and increased distribution frequency. All of these elements have tilted the power away from the service provider and towards the indirect channel and to the customer.

7.2 Evolving Routes to Market

Like all critical business processes, distribution channels have become more complex and more software intensive. The ability to customize channels to evolving customer locations and needs has created unprecedented information flows between sellers and buyers. For example, in telecommunications, as in every industry, the Internet has facilitated commerce by way of a new distribution channel, but it has complicated it through an old conflict, disintermediation.

Disintermediation occurs when a member of the distribution chain reaches around one or more channel partners to meet the customer. Normally, disintermediation occurs when a partner in the distribution network no longer adds value—meaning benefit beyond its cost—most often when the Internet's many capabilities have facilitated Web-based purchasing at no additional charges to the customer. In the most common manifestation of disintermediation, an Internet user can buy electronics, airline tickets or even an automobile from the source without a direct sales transaction in a store or showroom or even a distribution portal such as Amazon.com. Disintermediation is common when commodity products or services are also undifferentiated by distribution channel. While a customer seeking international calling services might invest in several visits to retailers to select a prepaid plan from a variety of programs, once a customer is committed to the plan, the same customer might refresh the account value by telephone or on the Internet.

Web-based self-provisioning, billing and account management is already eliminating some members of the distribution channel within the service provider's own employee base. Fewer or smaller billing centers result when many of these transactions are simplified through the availability of alternative channels.

In most industries, disintermediation arose when distributors were squeezed out of the chain. Oddly, disintermediation outside the service provider's enterprise occurred in the telecommunications services market in an opposite manner. A retail role inserted itself into the distribution channel when deregulation took place. Because nearly all facilities-based service providers operate in wholesale and retail markets, and because resellers need suppliers, disintermediation is common. The reseller buys wholesale service from the service provider that competes with the reseller in retail markets. A restructuring in which suppliers serve both distributors and the distributor's customers often results in channel conflict and other market disparities that service providers and their distribution partners need to reconcile.

Opportunities abound to differentiate within the distribution channel, so disintermediation is not destiny. The entire notion of value-added resale implies that the distributor adds something to the process that the customer wants. For technology products and services, the sales and delivery functions might require specialized presales technical support, assistance in installation or customization of the services, or providing financing or transaction support. The newly coined term re-intermediation describes the new role of the intermediary as a partner in the distribution process, as long as the customer sees value in the intermediary's presence. Part of the movement towards re-intermediation came about because technology replaced some of the traditional indirect sales channels, and self-service purchasing replaced expensive sales labor. Service providers passed along the savings to their customers, who became self-sufficient purchasers for most goods and services and demanded value and efficiency in the buying experience. Technology and self-service can replace most, but not all of the functions formerly accomplished by intermediaries. Re-intermediaries filled the gaps in the channel, especially in specialized applications.

Distribution channels are becoming more agile, especially in response to the reduction in the technology life cycle. A distribution infrastructure needs to be in place and ready to launch new services at increasing speed. Because short-lived technologies are most profitable in the beginning of the life cycle and priced close to cost at their decline, distribution is key to achieving the maximum profit in the short window of profitability. One method to leverage distribution reach is to structure the channel as a network rather than as an assembly line. Though outsourcing is not the requirement for large service providers that it is for new entrants to a market, a network of outside distributors can maximize market penetration and preserve in-house resources for key accounts. New entrants need to optimize their limited resources, and established service providers can gain flexibility and effectiveness by deploying intermediaries.

The Internet is not the only new route to market, though it garners the lion's share of media attention. Call centers have become sophisticated beyond their roots as answering services and fulfillment centers. Partnerships and alliances with complementary service providers and agencies have introduced value into the channel. Wireless application protocol (WAP)-enabled mobile phones promise to offer an innovative customization of direct marketing. Either by subscription or by event, service providers will be able to notify prospects and customers of messages of interest to the customer. While issues of privacy abound, a customer with a WAP-enabled phone can learn of promotions when—or where—the customer is most likely to buy. A prepaid wireless service provider might send a notification to the customer that the remaining time is beneath a customer-designated reorder point, and that an agency to refresh the account is nearby and open. Interactive television, while only an emerging technology, promises to become another distribution venue. Customers might sign up for services in response to a commercial or click on an item of interest from a product placement in a movie or other program. At the same time that distribution is ubiquitous, the targeting and execution of the associated advertisement will be so defined that the placements and messages will nearly be unobtrusive to those not seeking to buy. If not, customers will find that even their escapist entertainment venue is polluted with spam. An innovation in bricks-and-mortar distribution, the store-inside-a-store enables a service provider to open a retail presence without the commitment of leases or the expense of staff. T-Mobile placed Wi-Fi hotspots in Starbucks coffeehouses and Kinko's office services centers, enabling the wireless service provider to use retailers as their agents. Verizon Wireless placed self-service kiosks in the stores of electronics retailer Circuit City, through which customers can learn about its service offerings, then purchase and activate services through a store sales associate.

Payphones in public areas once offered local service providers distribution points for revenue-producing services, but the proliferation of customer-focused competitors and the penetration of mobile telephony squeezed public telecommunications margins until service providers needed to review their strategies. Some chose to use the units as distant distribution channels for services beyond traditional telephony. British Telecom replaced many of its old payphones with Internet-enabled phone terminals, where customers can send video e-mail, experiment with broadband service to surf the Internet, download wireless ringtones, and make conventional phone calls. Verizon used its payphones in

Manhattan to host Wi-Fi hotspots for its on-line customers. Other service providers, like BellSouth, chose to exit the market and focus on other distribution channels.

7.3 Cross-Channel Strategies

Telecommunications service providers use cross-channel strategies to meet customer demands, including Web-based billing, in-store handset purchases, and telephone-based customer care. Success across channels requires a common interface between customer contacts, to ensure the fulfillment of customer requests and to present an accurate account view to the customer. This seamlessness is supported by real-time updates of databases that are viewable by the customer, so that, for example, a customer account review on the Internet would reflect the newest purchases of service, the status of trouble-reporting repair requests, or changes to the customer's profile. All of the channels must present a similar appearance to the customer, facilitated by familiar colors, logos, and advertising materials.

Cross-channel selling distinguishes itself from multichannel selling through the integration of a variety of sales channels. It is one thing for a service provider to sell through several channels, and another to add value to the distribution chain through cross-channel strategies. Integrating sales channels enables the service provider to migrate sales from one channel to another to achieve the most cost-effective delivery. Thus, in the service life cycle, direct sales might be necessary, using technically knowledgeable representatives to convince technically knowledgeable buyers to purchase, for example, mobile broadband services. As the critical mass of buyers develops, the audience for the service has had sufficient exposure to the service to make purchase decisions based on their own research, and by then the service might well be less complex to configure and use. Migrating from expensive channels to more efficient delivery channels will also reduce costs and prices, paving the way for growth by exploiting the elasticity inherent in most telecommunications services.

Channel optimization refers to the process of moving customers to lower-priced channels to improve return on investment. Figure 7.1 demonstrates the extent that channel costs can be minimized when customers are reached economically. The least expensive method to reach customers, Web search, has an additional benefit of being much less intrusive than most other sales channels. Through the Internet search channel, the customer seeks out the service provider and controls the sales process. It is a desirable relationship from the point of view of the customer and the service provider.

Optimizing the delivery channel involves moving customer interactions from the more expensive methods to the more cost-effective channels, assuming that the leads produced from each method are of equal value. One local exchange service provider diverted 5% of its customers to lower-cost channels, saving $15 million in expenses and adding $40 million in revenues [1]. AT&T credited its click-to-chat Web site capability for tripling on-line sales in 6 months. ADC Telecom's Web-based customer service decreased call-center volumes by 25% and improved customer satisfaction. Bell Canada's Web site improvements resulted in diverting more than a third of its customer contact to on-line support.

Monopolist telecommunications service providers might once have begrudged losing control over the entire distribution process to intermediaries. Monopolists might resent the associated revenue sharing that accompanied the use of outside agents in the sales process. In an industry characterized by intense competition mitigated by significant growth, even the most reluctant incumbent service providers recognize the benefit they receive from leveraging their sales efforts. Qwest stated in 2001 that it expected to drive 50% of its sales through indirect channels [2]. One-quarter of Sprint's mobile PCS and Internet services sales derived from its retail partner Radio Shack. New entrants to the telecommunications services market did not boast the footprint of the market leaders, and some have specifically targeted second and third-tier markets that might be overlooked by the giants. New Edge Networks, a DSL provider targeting small businesses in hundreds of markets, relies on indirect channels, including partnerships with other DSL providers and hundreds of wholesale ISP providers. For service providers without an existing sales force, indirect channels create a wide coverage while tying expenses to revenues and responding to variations in the growth profile of the marketplace.

Qwest, for example, could participate in multichannel sales efforts by acting in the sales background for consultants and systems integrators recommending network-based solutions for their clients, hoping to be selected as the service provider. Instead, Qwest developed a cross-channel Business Partner Program to support nontraditional sales channels for its Application Service Provider (ASP) portfolio. The program offers sales support, customer relationship management, cobranding, and commissions to its partners. The program represents a shift from a product-sales orientation to a customer-centered solutions strategy.

Service providers who operate in multiple retail channels need to eliminate barriers between them, especially because multichannel customers are among the most desirable to serve. According to retail consultant J.C. Williams Group, cross-channel shoppers spend 50% more and visit stores 70% more frequently than the average shopper. To succeed in a cross-channel environment, the service provider needs to ensure that every component of the customer experience is seamless. Services described by a call-center sales representative must match exactly in features and price with those advertised on the Internet, barring explicitly labeled Internet-only promotions, because the telephone customer

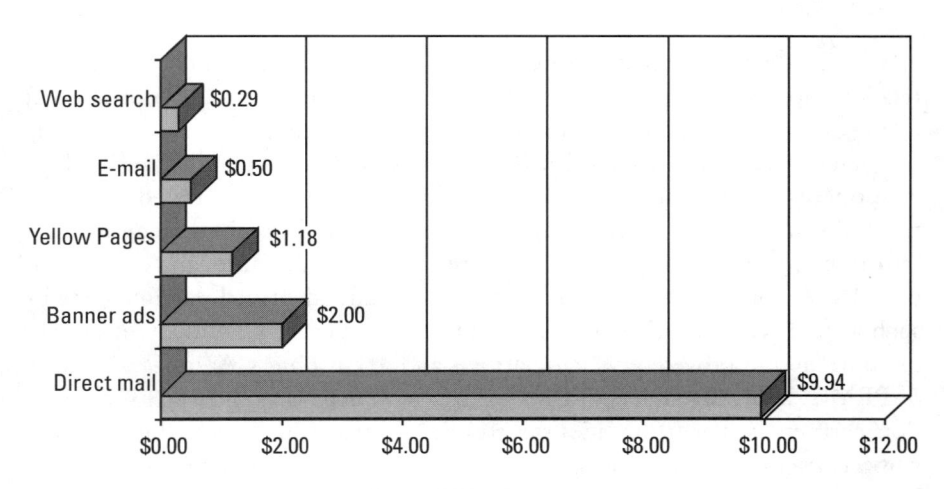

Figure 7.1 Lead costs per channel. (*After:* [3].)

Channel	Cost
Web search	$0.29
E-mail	$0.50
Yellow Pages	$1.18
Banner ads	$2.00
Direct mail	$9.94

might well be watching a screen during the conversation with the sales representative. In the back office, customer service, including problem resolution, loyalty programs, and promotions should be independent of the channel through which the customer enters the enterprise. To the degree feasible, each channel's representative should access and update the same databases.

7.4 Partner Relationship Management Systems

Partner relationship management (PRM) information systems, also known as channel relationship management systems, enable service providers to manage their complex relationships with distribution partners. Selling through indirect channels is economical but complex, and the new breed of information support emerged to enable disparate members of the same distribution chain to maintain communication and accuracy. Automation of channel management coordinates information exchange among enterprise functions such as marketing, sales, and customer care, whether they or not they reside under the same corporate umbrella. Partner relationship management systems serve a similar function to customer relationship management systems, by consolidating a variety of data points gleaned from transactions and core marketing processes in the interest of improving key business relationships. The comparison with CRM systems is apt, because indirect sales partners, like customers and unlike employees, are volunteers in the business relationship. Furthermore, the proliferation of PRM installations created its own competitive parity; the highest quality agents and indirect sales representatives require a certain level of infrastructure before they will represent a service provider.

According to market researcher META Group, the market for channel relationship management systems will exceed $550 million in 2005 or 2006 at its present growth rate of 15% to 20%. The growth is apparently warranted by the benefits accruing from a PRM system as compared to other technology investments. A December 2002 Gartner Group survey demonstrated that two-thirds of survey respondents believed that they had received demonstrable return on investment in PRM, as compared to CRM sales suites and other automated sales applications. The Gartner Group study identified tangible benefits such as reduced channel partner ramp-up, reduced order placement and tracking time, and improved service to partners and to customers. Another advantage of PRM systems for large service providers with extended and varied distribution partners is that it provides structure and consistency across the sales network. Participants in the sales organization can maintain contact through a Web interface, update order databases, and seek sales support.

Cable & Wireless augmented its existing back-office systems with a partner relationship management solution to assist in the transition from a direct sales force to indirect channel partners. The system, among its other capabilities, classified 25,000 active sales leads and set tracking alerts to ensure that partners were following through with their commitments. If a partner failed to meet expectations, the system reassigned the leads to other partners. Greece's mobile services provider COSMOTE deployed a PRM system to manage business agreements between carriers, bundle new services, and manage new service introductions. Qwest credited its PRM system for improving its lead close rate by 1%, reallocating $10 million in staff expenses, and improving channel sales by 10%. Broadband service provider Covad leverages a PRM system to create a fast-growing alliance program to position the service provider to compete with its large competitors.

Security is vitally important in PRM systems, because outsiders, though trusted, can access the corporate database. PRM systems use extranets, which offer a secure portal to the service provider's applications. The systems offer two forms of security: excluding access to all but channel partners, and excluding even the channel partners from unauthorized access to enterprise applications.

7.5 E-Business Channel Management

The Internet is an indispensable component of cross-channel marketing strategies. For channel partners, it is a repository of product information, status reporting, and support. For customers and potential customers, the service provider's Web site can create excitement about— and deliver—new services, perform account management, automate provisioning and trouble-reporting functions, provide extensive customer care, offer up-selling opportunities, and create customer intimacy, which leads to customer satisfaction and loyalty.

The first challenge is to get customers to the Web site. It is useful but not sufficient to register the site with the major search engines, because the quest for customer intimacy requires more than a single interaction. Some of the major service providers offer predictable services on their home pages that are duplicates of those offered on popular portals such as Yahoo!, such as reverse phone number lookups and Yellow Pages listings. While these are conveniences and fit with the core service provider business, they are no competition for the dedicated portals and would not draw customers on their own merits. On-line billing, available from most service providers, encourages customers to visit at least once in each billing cycle, and Web-based e-mail accounts ensure frequent visits by existing customers. Verizon has offered on-line courses in business skills and computing and certifications in computing technologies, making space on the course portal for promotions of Verizon services. AT&T Wireless has maintained a link to a tool that helps the customer select a plan based on individual calling patterns. BellSouth has included a tool for customers to learn whether or when broadband DSL service would be available to them.

Through technology, service providers can, within the bounds of privacy, develop intimacy with their customers. In the same way that a supermarket might print coupons for products that are complementary to those in a customer's basket and on-line bookseller Amazon.com recommends books based upon previous customer purchases and the customer's assessment of books already read, telecommunications service providers can tailor promotions to the actual usage of their customers. Qwest uses whatever channel the customer has selected—the call center, the Internet, or automated voice response to customize the interaction. When a Qwest customer calls for repair, the service provider polls its customer database to retrieve the customer's profile after the customer has entered the number into the voice response system. The options presented to the customer reflect the services that the customer uses, expediting the connection to the proper problem solver in the call center, and reducing the data entry requirements for the customer.

Telecom New Zealand opted to draw customer traffic to its Web site, because it viewed the Internet as the best channel to deliver detailed marketing information economically, but its ordinary catalogue engine was insufficient to handle the increased demands of the service provider. A content management system automated the process of developing Web pages, including the insertion of appropriate links on each page through a specified algorithm. The content management application standardized the template, creating a consistent and approved face to the customer, and the system enabled Telecom New Zealand to add tools to its site to tailor its features to the user's objectives. The service provider attributed a doubling or more of visits to the site to the new look and capabilities of the content management implementation.

Turkish Internet service provider Superonline uses content management tools to translate English-language Web sites—which make up the vast majority of Internet pages—into Turkish for its customers, creating a reason for its customers to return to its portal frequently. BellSouth uses a content management system to manage thousands of service contracts with its wholesale customers. Sprint PCS used content management systems to offer dozens of customized sites tailored to the unique needs of its business-to-business wireless customers.

7.6 Strategic Alliances

Public relations professionals fill the industry trade press with exuberant headlines announcing strategic alliances among telecommunications service providers and their partners; however, only some of these represent genuine strategic alliances. Many of the new relationships reported as strategic alliances simply describe conventional roles in the distribution chain between telecommunications services suppliers and resellers or agents, or between service providers and their own suppliers. When a medium-sized software developer or consulting firm issues a breathless press release announcing an alliance with a major service provider, one test of its validity is to search the service provider's Web site for an announcement mirroring the partner's excitement. If validation is absent, which is most often the case, then the supposed alliance is probably just an overenthusiastic euphemism for a sale. The test is whether the alliance crosses the industry horizontally or vertically. Most truly strategic alliances take place horizontally on the distribution chain, and almost always between two partners who exercise similar power in the marketplace.

To simplify the terminology, while it is true that all marketing agreements between members of a distribution chain are strategic, and all are alliances, strategic alliances are something more than that. A strategic alliance joins two horizontally aligned channel members who are either competitors or strategic potential competitors in a partnership to assume a significant presence upon the same position of the distribution channel. The British Telecom/AT&T Concert venture was a strategic alliance, although it did not survive, its failure putting that initiative in the company of more than half of all strategic alliances. The cobranding of Cingular wireless services by its owners SBC and BellSouth is a strategic alliance. The efforts of Yahoo! and BT Openworld, the ISP division of British Telecom, to cobrand a portal is another example of a strategic alliance. The Concert and Cingular initiatives are examples of market leaders in adjacent geographical markets expanding their footprints. These alliances are especially powerful if each partner already has a strong brand presence in its own territorial or horizontal market. Indeed, a relationship in which one partner's brand is strong and the other is weak is not an alliance; it is mentoring.

The Yahoo!/British Telecom was an effort to leverage the strengths of its two horizontally linked market members to benefit the combination. Yahoo! brought portal experience and leadership that would encourage United Kingdom customers to get or stay on-line, which benefits British Telecom. BT brought a customer base of narrowband and broadband users to the Yahoo! portal, personalized to cultural uniqueness and individual needs. For example, its radio service feature anticipates the user's music preferences.

Though it is widely known that about 60% of strategic alliances fail [4], the pace of alliance building continues. The median life span of a strategic alliance is 7 years, and about 80% of joint ventures end in a sale by one of the partners [5]. The end of the GlobalOne alliance, a consortium between Deutsche Telekom, France Télécom, and Sprint dissolved while Sprint was in merger negotiations with MCI WorldCom. The strategic alliance can be a precursor to a merger, as the cobranding of Bell Atlantic and NYNEX wireless services led to their merger. A strategic alliance gives the participants an opportunity to test the same qualities and fit that are necessary to a successful merger. Alliances represent the cohabitation to a merger's marriage, and a failed alliance is easier to unravel than an unsuccessful merger. Alliances do bear a burden that is reduced for merger partners: management needs to strike a balance between protecting corporate information and providing the partner with sufficient data to succeed. As the venture partner is most often a competitor, the temptation to hide valuable knowledge is strong. Several factors increase the likelihood of success for a strategic alliance. A successful strategic alliance will provide value to both partners that would have been unavailable without the alliance in place. Additive benefits do not invigorate the market; innovative combinations do. Successful alliances have clear boundaries and unquestioned management commitment. As with cohabitation, strategic alliances do not have to end in a merger, but they will be more successful if all of the parties treat it as a going concern with no foreseeable end date.

7.7 Scoping the Channel as a Customer-Centered Strategy

From time to time, the debate over whether facilities-based service providers should operate in the retail channel reemerges. Occasionally there is a call for regulatory intervention to separate the wholesale and retail service provider functions. The main concern is that channel conflict would prevent the service providers from treating resellers with the same enthusiasm as other customer segments. Channel conflict occurs when two participants in line along the distribution channel compete for the same customer, for example, an incumbent telecommunications service provider that offers broadband service and a DSL provider enhancing and reselling the service provider's facility. The traditional remedy for channel conflict is through explicit contracts or the legal system. A different remedy involves a reassessment of channel strategy.

Figure 7.2 depicts four types of telecommunications services customers by their reason for buying services, their unique network need, and by marketing approach. The value-added reseller adds capability to the network, either for vertical markets or to differentiate the service. The resale marketer buys telecommunications services at a discount and then markets and distributes it. On the bottom row, the enterprise network customer and the retail customer, defined as consumers and small-business customers, purchase services for their own use or for the use of customers they represent. Because these customers use rather than resell the service, the telecommunications service provider needs to meet their unique needs through targeting and segmentation. Both members of the top row of the chart, the value-added reseller and the resale marketer, purchase services for enhancement and resale, and they both represent a source of channel conflict when they compete for the customers on the bottom row of the chart.

Along a single dimension of the chart, the top row versus the bottom row, it appears that the top row represents the traditional wholesale market, those customers that will enhance and sell the facilities offered by the service provider. Using similar logic, the bottom row represents the conventional retail market. Nonetheless, the chart also identifies one dimension upon which the diagram is best split vertically: the need for features in the network. In this measure, the enterprise network, acting in some ways as a local service provider, requires the same network management tools as any local service provider. Indeed, the competitors to serve the enterprise network customer include the enterprise itself, choosing to bring telecommunications services in-house because its own service provider, distracted by a plethora of disparate channels, fails to meet its needs. While some enterprises that support internal networks do so because of

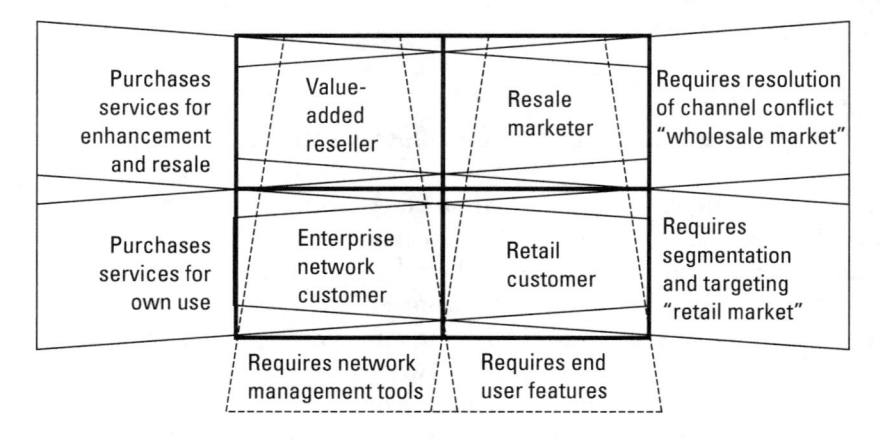

Figure 7.2 Channel strategies by customer type.

mission-critical functions or unique needs, others would as soon outsource the function to a trusted partner. Similarly, for user-friendly features, the retail customer resembles the resale marketer more than the enterprise network customer. Thus a strategic decision might streamline the channels for a facilities-based service provider that chooses to serve customers operating their own networks and requiring service quality commitments, network management tools, and other capabilities the facilities-based service provider can offer. Targeting only the right column of the table shifts the channel conflict to the resale marketer, but a different strategic approach can serve these channels, simply by changing the role of the resale marketer from customer to indirect channel partner. Sometimes competitive threats are strongest outside the channels completely. For example, VoIP, which threatens to revolutionize the infrastructure of the network, is likely to change the entire value proposition to the resale and retail markets.

References

[1] Grieve, Kevin, and Edgar Ortiz, "Customer Value Management Sins Can Be Costly," *DM Review*, Vol. 13, No. 11, 2003, pp. 38–40.

[2] Lambert, Peter, "All For One: Partner Management Makes Its Pitch in Network Services," *PHONE+*, Vol. 16, No. 6, 2002.

[3] Bluestreak, http://www.bluestreak.com/web/.

[4] Caroline Ellis, "Making Strategic Alliances Succeed," *Harvard Business Review*, Vol. 74, No. 4, 1996, pp. 8–9.

[5] Bleeke, Joel, and David Ernst, "Is Your Strategic Alliance Really a Sale?" *Harvard Business Review*, Vol. 73, No. 1, 1995, pp. 97–105.

8

Stakeholder Management

8.1 The Importance of Stakeholders

The service provider's influence reaches out well beyond the boundaries of the enterprise, and well beyond its ordinary sphere of contact with its distribution chain. Those individuals or entities with a stake in the operations or other activities of the service provider are called stakeholders, an apt if unimaginative title. Stakeholders, in individual ways, share in the success or failure of the enterprise, sometimes directly, sometimes indirectly, and sometimes inversely. As one example, competitors do all three. They benefit indirectly from innovations brought to market by others when they can imitate them, and they profit directly when lobbying paid for by others brings about a favorable regulatory outcome. Finally, competitors benefit inversely when others exit the market for reasons unrelated to the strength of the overall market, for example, an unrecoverable corporate accounting scandal limited only to the perpetrator. A stake can be based upon ownership, such as for shareholders or participants in joint ventures. Some stakeholders have a legal stake outside of ownership, such as a contract to receive services for a specified timeframe or under specific conditions. Still others might simply claim an interest in the activities of the service provider, such as environmental groups representing the use of natural resources, human rights groups conducting employee advocacy, or consumer groups.

Though it appears paradoxical, the needs of stakeholders are especially important to the customer-centered enterprise, even when the stakeholders do not represent customers or specified customer segments. Stakeholders affect customers for two reasons: first, because they shape the business—by regulating the relationship the service provider has with customers as governmental entities do, by affecting the service portfolio or the customer relationship as members of the distribution chain do, by affecting the brand image held by the customer as the community does, or by nurturing the capitalization soundness of the enterprise as financial stakeholders do. Furthermore, stakeholders and customers are not mutually exclusive groups; the fervent environmentalist might evaluate a service provider's philanthropic profile or ecological history before selecting a service provider, and the technical contact representing the buyer within a large corporate account might well participate actively in an important user group. Figure 8.1 depicts the swarm of stakeholders surrounding the telecommunications service provider. Part of the value that the service provider takes to its owners and its employees includes its relationships with diverse outsiders, and nurturing the business environment through stakeholder relationships is one of the tasks of the marketing function.

Stakeholder theory has gained importance in the recent past and has raised outsiders' expectations about corporate performance, not just its profitability. Corporate social responsibility is no longer only good manners or good morals; it is good business. Around-the-clock news media seeking stories and the proliferation of opinions and gossip via the Internet have made all corporate behavior more visible than ever. Some mutual fund managers seek socially responsible investments, to satisfy the investment needs of what was once a fringe investor group but is now an important social movement. A service provider might be vulnerable to criticism and harm to its image and investment attractiveness for conducting business operations with a fixated orientation to profitability and

The Stakeholder Context for Telecommunications Service Providers

Community
Consumer groups
Public interest groups
Opinion leaders
Philanthropy
Mainstream media

Customer

Distribution chain
Employees
Suppliers
Contractors
Distribution partners

Service provider

Legal
Industry regulators
Business regulators
Unions

Financial
Shareholders
Board of directors
Financial institutions
Industry analysts

Indirect distribution chain
Competitors
Customer trade associations
Supplier trade associations
User groups

Figure 8.1 The stakeholder context.

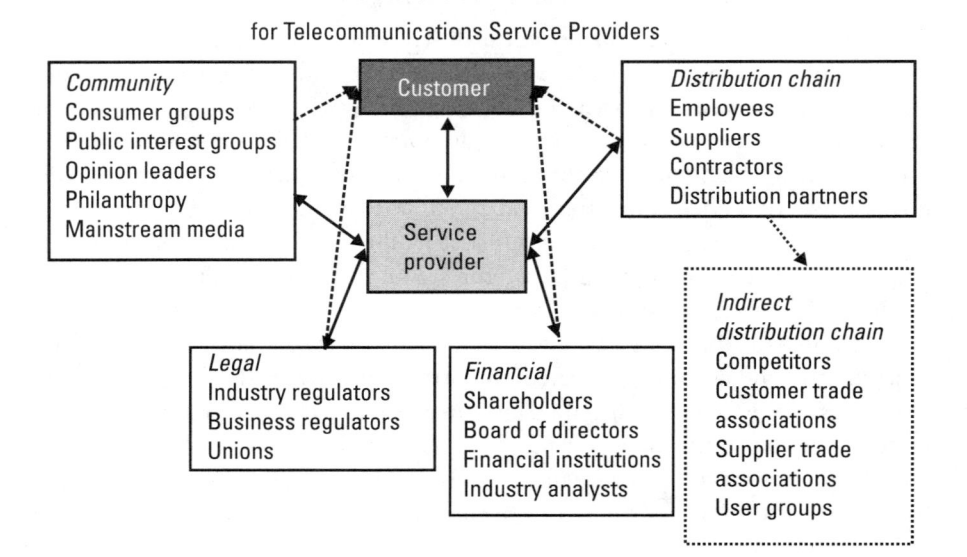

insensitivity to social responsibility. Globalization means that service providers with operations in countries with delicate governmental structures might need to include a wide assortment of variables in their cost-benefit analyses.

Dow Jones supports a series of sustainability indexes to rank corporate performance against social responsibility criteria. Of the seven telecommunications service providers represented favorably on the DJSI World 2003/2004 index, none were U.S.-based. Dow Jones also maintains the Dow Jones STOXX index for European companies. The FTSE4Good index in the United Kingdom serves a similar purpose. In 2004, its GoodUS index included AT&T, Comcast, Level 3 Communications, Qwest, SBC, Time Warner, and Verizon. Both of these powerful indexing companies claim that sustainability, that is, long-term corporate viability, improves when corporations conduct socially responsible practices.

British Telecom inaugurated its Better World program to promote its corporate values regarding society and the environment and maintains a Web site to publicize its commitment. The site provides policy statements, press releases announcing community activities, white papers, surveys, and even a game that lets the user make the difficult ethical decisions facing large corporations. The service provider's efforts earned it the top position for the telecommunications sector in the Dow Jones sustainability index of European companies for its economic, environmental, and social performance.

Service providers producing environmental reports include Norway's Telenor, Swisscom, and Spain's mobile service provider Telefonica Moviles. Deutsche Telekom maintains a waste management database. Demonstrating that it practiced what it preached, Swedish telecommunications service provider Telia issued its lengthy environmental reports only in electronic form.

Stakeholder management refers to the development and implementation of policies and practices that take into account the goals and concerns of all relevant stakeholders [1]. The process is straightforward and practical, outlined in Figure 8.2.

The first step identifies the universe of stakeholders, segmenting them into groups if necessary. Stakeholder management involves education—learning what the stakeholder concerns are—and making an assessment to put each concern in context. One method for prioritizing stakeholder concerns is to evaluate their salience, specifically their power, legitimacy, and urgency [2]. Determining salience is purely a judgment call and stems from the presence and strength of each of the three attributes.

Stakeholders that hold power hold priority over those whose unmet concerns would have only a modest impact on the success of the enterprise. Power implies the ability to influence others or control strategic resources. A stakeholder who can control the service provider's environment—a regulator or a sole supplier of strategic technology—has more power than a consumer, no matter

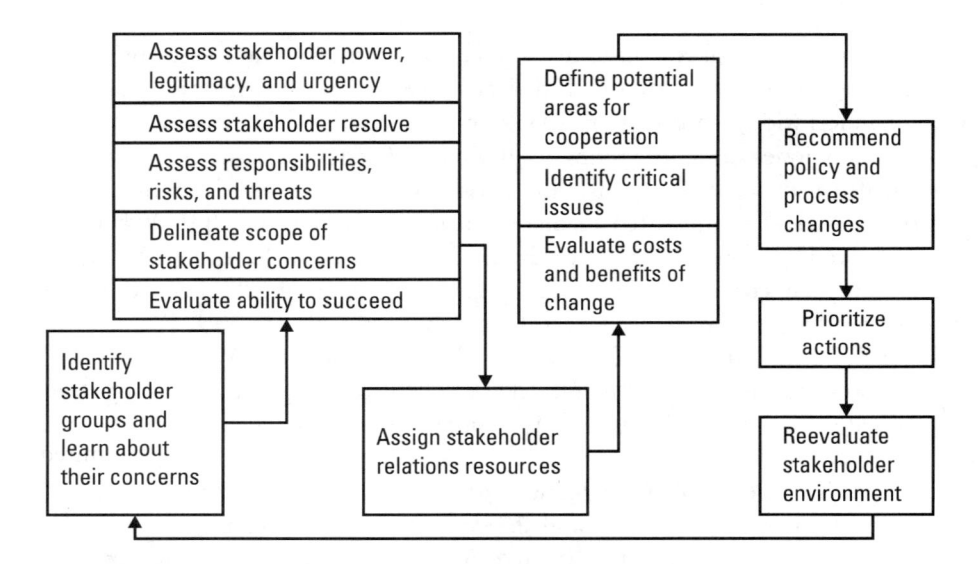

Figure 8.2 The stakeholder management process.

how noble the consumer's cause. Grassroots organizations and unions are examples of stakeholders whose power emanates from their unification. Power is relative; power exists when the stakeholder can force the enterprise to do something it would not have chosen to do otherwise.

Power on its own is normally not sufficient to give a stakeholder priority. The powerful stakeholder who holds legitimacy or urgency will gain priority over the stakeholder with power alone. Stakeholders have legitimacy when others in the society outside the context of the extended enterprise view their claims as reasonable or appropriate. Urgency refers to whether the claim is time-critical and important. Urgent claims without legitimacy earn a low priority; legitimate claims without urgency do not require immediate attention, subject to the stakeholder's level of resolve.

A combination of power and resolve, even if the stakeholder's objectives hold little legitimacy or urgency as judged by the service provider, might gain visibility to the stake. Such stakeholders, if perceived as potentially dangerous, pose a priority challenge that might very well override a more reasoned analysis. Stakeholders with high power and high resolve are afforded a high priority; stakeholders with low interest, whatever their power, are less of a threat than those with high resolve and moderate power.

Responsibilities, risks, and threats affect the service provider's decision to apply resources to any expressed or potential stakeholder issue. Service providers can act preemptively, especially when competitors experience stakeholder setbacks. Sprint and WorldCom announced plans to spin off Internet backbone facilities in anticipation of regulatory concerns prior to their unconsummated merger in 2000. Two local service providers, Frontier Communications and SNET, announced plans to separate their wholesale and retail divisions in advance of an anticipated regulatory mandate. Soon, Global Crossing acquired Frontier, SBC acquired SNET, and the initiatives faded away, though the regulatory dispute still simmers. Local service provider Alltel and others in the industry replaced auditor Arthur Andersen when scandals surfaced at Global Crossing and WorldCom. Service providers might act preemptively to stave off potential future issues with stakeholder groups. Preemptive activity is desirable when the risk of future negative publicity or other costs is higher than the present value of making the change, in dollars or positive public relations.

The scope of stakeholder concerns affects their priority. Typically, stakeholder concerns involve the service provider taking action within a limited sphere of interest. Local regulators gain more attention when they raise service issues instead of human resources issues; environmentalists who concern themselves with the ecology will gain more attention than if they advocated a pricing restructure. The next test is the service provider's ability to succeed at resolving the issues raised by the stakeholder. If the potential for success is small, the risk of negative outcome is high even if the service provider attempts to address the stakeholder concern. All of these elements affect the service provider's assignment of its limited stakeholder relations resources. Indeed, to capture more attention, it is the job of stakeholders to maximize if not exaggerate the salience of their requests, by amassing resources to gain power, or to assert their legitimacy or urgency.

Stakeholders are not necessarily adversaries. Partnerships with stakeholders can improve the image of the enterprise in the community. The next phase of the process seeks opportunities for cooperation, to transform a potential opponent into a partner, by identifying the issues most critical to the stakeholder and assessing their impact to the enterprise in a conventional cost-benefit analysis. The rest of the process follows the path of virtually any corporate initiative, and leads back to reassessment of the stakeholder situation at the commencement of the process.

An approach called soft systems methodology or SSM arose several decades ago to help analysts discern the relationship between stakeholders and the complex problems they seek to resolve. Described by Checkland [3], the SSM methodology features a seven-step process that surpasses conventional computer systems analysis in its ability to resolve ill-defined problems. SSM begins with a comparison of the world as it is with the way it should be. After studying the present system, the researchers develop a model of an ideal system that would work better. The differences between the existing model and the ideal become the starting point for change. Applied regularly to the development of information technologies, this approach is useful for any situation where stakeholders are diverse and the challenges are complex. Other approaches that feature stakeholders prominently in their systems analysis techniques include systems dynamics, viable system diagnosis, strategic assumption surfacing and testing, interactive planning, and critical systems heuristics.

8.2 Customers as Stakeholders Throughout Changing Regulations

Service providers that began as monopolists found the regulatory environment to be their bane and their protector at alternate times. Service providers have crossed swords with each other, pitting incumbent local exchange providers or alternative access service providers against startup competitors, regional ex-monopolists expanding their territories, their own resale customers, and many other permutations of conflict. Typically, most participants who have the means, or who are represented by vocal trade groups, use all of the ammunition they can muster in the marketplace.

Local number portability in the United States is one example. The Telecommunications Act of 1996 mandated that wireless or wireline customers who switch service providers could keep their phone numbers (assuming that the customer remains in the region served by the original service provider). Wireless service providers sought to postpone the implementation and were granted several delays before the service became active in November 2003. In one document filed with the FCC, Alltel, AT&T Wireless, Cingular, Nextel, and Sprint requested several barriers to portability: switching fees, refusal to transfer services for users with unpaid balances, limiting the times of day that numbers can be switched, and refusing to switch users when the customer information does not match its records exactly, including abbreviations [4]. Verizon was conspicuously absent from this group of dissenters, though it had once been a vigorous opponent of local number portability. As a marketing strategy, Verizon was as correct to support local number portability as its opponents were to contest it. As the wireless industry leader—and recently anointed—Verizon's defensive strategy was perfect. The customers that left other service providers would come to Verizon as their first choice, and its own customers would experience noneconomic switching costs to leave the market leader. Market research demonstrating that Verizon was the first choice for churners was in the trade press and the company probably already owned proprietary research that came to the same conclusions. T-Mobile, experiencing growing market share, took the same position as did Verizon. In the end, the FCC's decision to include wireline-to-wireless portability, whether it represented a concession to wireless industry complaints, made the transition more palatable to wireless service providers by greatly expanding their market.

Fortunately for service providers, most customers are unaware of the backroom and courtroom arguments between service providers. A consumer group objected to the recovery charge imposed on wireless customers for funding the conversion to local number portability and particularly to the differences in amount between service providers and the unwillingness of service providers to itemize what the charges were funding. A national newspaper headlined Verizon's plan to charge double the expected fee for funding its compliance to portability rules two days after the ruling took effect.

In a transforming market, service providers need to use every tool they have to position themselves. On the other hand, they need to balance the effort they place into investing in the regulatory process. There are few victories in transitional regulations that will make the difference between success and failure in a genuinely competitive marketplace. The resources that the service provider dedicates to regulatory conquest are not in the service of meeting customer needs. Transition to deregulation has already created several surrealistic contests between service providers and their own large customers. Regulatory management is a skill with a shortening shelf life, because customers expect a different kind of relationship with their vendors, and they will find it elsewhere on the telecommunications battlefield. Service providers need to reconcile their desire to manage the regulatory process against the potential backlash they will experience from their customers.

8.3 Issues Management

Simply stated, an issue is any matter about which the involved parties disagree about its disposition. The stakeholder function of monitoring public policy or other issues of strategic importance to the successful continuation of the enterprise is called *issues management*. While it might once have been acceptable under less competitive markets for service providers to react to issues as they arose, today's telecommunications service provider must seek out the strategic issues before they are at hand. This requires an active issues management program that includes both data collection and rigorous analysis.

These issues might involve public policy, such as privacy. Telecommunications service providers in the United States discovered that they were embroiled in a well-publicized dispute with the Recording Industry of America when they were unwilling to disclose the identity of customers who were illegally downloading copyrighted material on the Internet. Privacy and telecommunications was already an explosive issue, with customers providing sensitive information to Internet vendors and on-line healthcare providers and vendors tracking Internet surfers' movements. Enabling customers to opt in and out of advertising by keeping phone numbers and e-mail addresses private affects access and transport providers, complicated by unscrupulous marketers who harvested active addresses from the very opt-out measures that users took.

Moreover, the privacy issue is gaining ground in the wireless arena, because of the promise of location-sensitive advertising and intrusive camera phones. Local service providers have used privacy concerns to sell blocking services to customers who did not wish to receive unsolicited calls. That some customers are willing to pay a premium for location-based wireless services, including ads and coupons, while others are protective of their personal whereabouts highlights the technical and policy difficulties of privacy issues.

Issues management makes up at least most of the content of the stakeholder management process. The process is iterative; stakeholders probably raise most of the issues, but the emergence of unclaimed strategic issues will result in a hunt for their likely stakeholders. Some issues that arise are independent of the service provider or the obvious stakeholder. In the past, issues have emerged from catastrophic events, from literature (such as ecology or safety), from corporate scandal, and from politics.

A proactive approach to issues management has many benefits. The service provider will be more successful if it can frame the debate of strategically important issues early and on its own terms. The service provider can practice diplomacy and negotiations more successfully before the debate has accrued a public and bitter history. Early assessment of issues enables the service provider to rank them according to probability of occurrence and potential for damage, though gauging probabilities is an inexact science at best. Even with conscientious effort on the service provider's part, issues management is a slight misnomer, as it is the nature of issues in the first place to manage the service provider and not the other way around. It is a more accurate interpretation to assume that what the service provider is managing is not the issue itself but its own response to the issue's ramifications.

The life cycle of an issue graphically resembles the diffusion of innovation chart (see Figure 1.1 in Chapter 1), a nearly bell-shaped curve that gains momentum, achieves critical mass, and then trails off into a decline, either because it has been resolved or because it has lost its importance or relevance. In its most controlled manifestation, the issues management process begins with a function called environmental scanning, a somewhat self-important depiction of a task incumbent upon all professionals: to read the trade press and stay informed about the industry and related strategic industries.

The environmental scanner looks beyond the issue on its face, to see the possible opportunities and threats beyond the obvious. Who might have imagined that the first successful camera phone would be a wireless one? What might have guessed that the killer application that made the Internet indispensable enough to penetrate nearly the entire population of the United States was different for each user and varied by month? According to market researcher

Nielsen//NetRatings, at the beginning of 2004, the largest e-commerce site in Europe was eBay. The environmental scanner that spots the trend before anyone else does will earn the first-mover advantage for opportunities and will enjoy a period of advance notification to respond to threats before the competition.

Some observers make the distinction between routine issues, which significantly affect the business in a favorable or unfavorable way, and strategic issues, which cut to the fundamental business model. Broadband delivery of entertainment services has long been predicted to invalidate the in-store video-rental business model, yet to date the change in media from videotape to DVD has had a more substantial impact to video-rental retailers than more than a decade of imminent broadband services. Furthermore, video-rental stores have diversified to complementary products to retain market share against the comfort of the couch: new and used DVD sales, magazines, and popcorn, all products which have not yet found a download market.

Strategic issues have far-reaching implications, which might be well beyond the scope of the conventional marketing function, and the data collection function and some of the analysis might reside in the business development department or strategic planning. Marketing management is keenly interested in the conclusions of the formal environmental scanning effort, and should apply resources to assessing the likely impact upon customer relationships, future marketing efforts, and potential targeting and revenue enhancement opportunities. The concerns of present-day issues management have a shorter timeframe for resolution, and missteps are more visible and more consequential.

The issues management process covers similar terrain to the stakeholder management process [5]. It begins with identifying the issues, then analyzing the most significant trends. The process moves to prioritizing the issues based on their importance, formulating a response or set of responses, putting the recommendations into action, and monitoring the results. Issues management, as compared to the stakeholder management process, retains one important difference of great consequence. When stakeholders bring issues to light, they often approach many of the market leaders and usually the press. Every competitor has access to the same information and a considerable amount of analysis, as posted in the general or trade press or as presented to arbitrators or regulators in a public forum. The issues unearthed through internal environmental scanning are secret, which adds to their value.

Cable service provider Comcast decided that theft of its service had reached crisis proportions. It established an amnesty program, in which guilty parties could turn themselves in without consequences. This program had two results. First, it recovered more than $2 million in revenues, and it served notice to persistent service bandits that there would be consequences for continued cable theft. There are many ways to resolve the issues identified by the issues management process, and service providers need to consider the long-term effects on customers.

8.4 Crisis Management

What makes a crisis a special management challenge is that by definition, it was unplanned. Most corporate crises come to light suddenly, but only a minority relate to a moment of disaster. Most crises incubate and are well known to management as potential calamities for some time. The external communications function of the marketing department places marketers in the center of crisis management. At such a critical time for the service provider, the success of the recovery from the crisis can be more dependent on the crisis communications than the resolution of the original problem. Furthermore, for service related problems, the telecommunications network is so integral to the functioning of society at large that excellent crisis handling can obviate future governmental intervention in business operations.

Crises have the potential to wreak significant impact on a brand, a corporate image, or profitability, and they threaten the viability of the organization as a going concern. Crises can have external causes. Verizon's post-September 11 efforts received an award for the service provider's contribution to disaster recovery and crisis management, a crisis that affected the service provider's own personnel and investment, but that undoubtedly had a more substantial impact on some of its customers. Indeed, service providers nearly always find themselves on hand for natural and environmental disasters, replacing infrastructure, providing temporary fixed-line and wireless services, and offering emergency communications to customers in the fray.

Internal crises might have their origins in the service provider's behavior, such as the WorldCom and Global Crossing scandals, or in the service provider's inaction, often because of social issues or human resources policies. The Institute for Crisis Management, a consulting and market research firm, placed the telecommunications industry at the top of its list of crisis-prone industries in 2002 and 2000, an index calculated from a content analysis of news stories.

Recent research has identified four building blocks that distinguish the crisis-prone industry from the crisis-prepared [6]. Described as the "onion model of crisis management" because the building blocks peel away like layers, the model describes four levels of crisis inclination. The first is individual, then organizational culture, structure, and strategies. The individual's inclination to crisis is at the core of the metaphorical onion. Thus the enterprise's proneness to crisis is management-led. Moreover, the behavior of crisis-prone management—such as defensiveness and denial—are the qualities that lead the organization into unmanageable misfortunes. Management also fosters its crisis-prone or crisis-prepared viewpoint throughout the organization, affecting the employees that need to carry out crisis management planning and handle crises as they occur. At the next higher level, the organizational structure can support or obstruct crisis management efforts. Last, the corporate strategies assist or impede crisis preparedness.

LinkLINE Communications, a small ISP in California, was one of many victims of an extortionist who had gained private credit information from linkLINE and other on-line vendors. Its crisis management plan was straightforward and effective. First, the company refrained from notifying the hacker that it had detected the intrusion while it monitored his activities with the assistance of law enforcement personnel. The ISP worked with its credit card partners to notify them of potential fraud, and gained their consent to protect the customers and the future partnership with the ISP. This relationship with financial institutions was undoubtedly more important to the ISP than to the banks; with linkLINE providing only 15,000 subscribers, the creditors might not have been willing to cooperate had linkLINE been less proactive. The ISP notified its customers about the crisis, issued a press release, and posted questions and answers on its Web site. It developed a training manual and trained its customer service representatives. One of the company founders received training to become the spokesperson, as did a backup spokesperson. The crisis management team put a program in place to notify other stakeholders. This service provider had the option to stop the security breach without a crisis communication program, and many other on-line sites continued to face the threat for a longer duration. Two weeks after the crisis, the ISP experienced no net loss of customers.

Crisis management professionals recommend that service providers establish a crisis plan and update it annually. Though it is impossible to predict the exact crisis that will transpire or when, it is safe to predict that a crisis of some sort will happen sometime. The plan can be specific about virtually everything except the particular crisis that will occur and the time it will burst forth. General plans can cover the departure of a key executive, an instance of workplace-related violence, a labor dispute, or a tragedy affecting employees or customers. The plan can describe the means to keep employees informed of unfolding events, which is both a moral obligation and pragmatically necessary to reduce the stress that could later affect worker performance. Customers comprise another crucial stakeholder group and often demonstrate great tolerance for interruptions in normal customer attention when they are part of the crisis management circle.

Though much of crisis management activity is reactive by its nature, like disaster recovery, media planning in advance of the crisis is essential. Without knowing the substance of the crisis, management can rehearse and prepare for those first media exposures and create plans for resolving the crisis and communicating the enterprise's activities effectively. For example, whistleblowers were at the heart of several recent corporate scandals. Typically, they approached media outlets as a last resort, out of frustration that their concerns went unheeded by management. The organization that sustains a clearinghouse for whistleblower concerns—assuming that it wants to eliminate improper behavior—will identify smoldering crises before they become untenable.

A trained spokesperson is a great asset when the media swarm occurs. Typically, the media arrive well before the enterprise has an opportunity to craft its formal statement. When a statement is released, reporters normally continue to ask questions beyond its scope. Furthermore, the spokesperson, often a high-ranking manager, has other responsibilities and might not be in the vicinity of corporate headquarters at the time of the crisis. Assigning and training understudies for the role of media contact is worthwhile.

The plan cannot assign all of the particular members of the crisis team, but it can specify the departmental representation for the team and define the process for selecting team members. A catastrophic network failure would require a different crisis team than would a frenzied picketing operation. No matter whether the normal organizational is highly decentralized, a crisis management team should not be a democracy; a clear chain of command is critical. The section describing the team organization should plan the communications among team members, including backup if the network is the problem. After the plan is complete, the likeliest team members should receive needed training and then participate in a crisis drill to evoke questions and possible additions to the plan. Uploading a generic version of the plan on the company intranet provides all employees with the opportunity to review it. For some, it will establish their disaster-related responsibilities. The rest of the employees will be confident that their employer has taken the effort to plan for the unthinkable.

WorldCom, in the midst of revelations of accounting irregularities and other corporate misdeeds leading to the largest U.S. bankruptcy until that time, launched an Internet microsite and posted news as it was made public. This had to be among the most difficult tasks to perform under the circumstances. The company posted its press releases, created links to court documents, and explained the arcane details of a Chapter 11 bankruptcy. Disseminating information during a crisis is one of the most unpleasant yet valuable stakeholder investments a service provider can make.

References

[1]   Post, James E., Lee E. Preston, and Sybille Sachs, *Redefining the Corporation: Stakeholder Management and Organizational Wealth*, Stanford, CA: Stanford University Press, 2002.

[2] Mitchell, Ronald K., Bradley R. Agle, and Donna J. Wood, "Toward a Theory of Stakeholder Identification and Salience: Defining the Principle of Who and What Really Counts," *Academy of Management Review*, Vol. 22, No. 4, 1997, pp. 853–886.

[3] Checkland, P., *Systems Thinking, Systems Practice,* New York: Wiley, 1981.

[4] Pappalardo, Denise, "Portability Near for Wireless Set," *Network World*, Vol. 20, No. 34, 2003.

[5] Carroll, Archie B., and Ann K. Buchholtz, *Business and Society: Ethics and Stakeholder Management*, Cincinnati, OH: South-Western, 1999.

[6] Pauchant, Thierry C., and Ian I. Mitroff, *Transforming the Crisis-Prone Organization*, San Francisco, CA: Jossey-Bass, 1992.

9

Market Segmentation

9.1 Markets and Submarkets

A competitive telecommunications marketplace demands that service providers offer a variety of alternatives to meet the diverse needs of customers. Market segmentation might once have been a matter of drawing a line between businesses and consumers, or high usage and low usage, but not anymore. Today's segmentation techniques benefit the customer and the service provider, by creating customized service portfolios targeted to attractive market segments.

At the very least, the service provider can begin with the broad view of potential segments or submarkets. The most apparent distinction between customers is geographic, and under monopoly conditions, geographic segmentation translated into the mandates of the relevant regulatory jurisdiction, which represented customer desires indirectly, broadly and disproportionately protectively. Today's mosaic of regulatory residue requires national and global service providers to monitor and comply with regulations in nearly all stages of deregulation as it appears around the world and across the marketplace.

Geographic segmentation also includes the variable of density, which affects both the cost of offering services and, in the inverse, the need for access to the wider network. The reality of geography, while nearly irrelevant in the transport arena due to the globe-shrinking Internet, becomes crucial from the perspective of access. Customers in sparsely populated areas, especially consumers, who have the least access to information and transactions in their remote localities, are punished with high access costs without the critical mass of customers to divide fixed costs and lower the averages. This disparity in access is the main reason for regulatory fretting and continued market involvement in the United States and elsewhere.

Refining the model slightly, the service provider could target a demographic group based on age, nationality, or geography, a psychographic group on the basis of beliefs, attitudes, or interests, or a behavior group based on usage patterns. The list of traditional segmentation strategies, always shown in the same sequence, also represented a kind of evolution in segmentation strategies. Figure 9.1 depicts the growth in complexity and effectiveness of segmenting techniques.

Demographic differences do affect buying behavior, and when all telecommunications markets were still monopolies, a slightly targeted service might have gained the acceptance of regulators and customers. But competition brings portfolio variety, supported by information technology in service development, operations, and customer profiling. If the service provider stops at broad measures, it will offer an undifferentiated portfolio to uninterested customers. Successfully competitive service providers do not control the portfolio, customers do. Today, coarse distinctions between markets represent only a starting point on the segmentation journey.

Psychographic variables reside higher on the evolutionary scale because these lifestyle characteristics can predict customer behavior. Even so, these variables mean very little in some of the most important customer segments, such as large business and small-to-medium business. For those markets, vertical marketing strategies are more effective. A further flaw in psychographic segmentation is that it predicts customer behavior on the basis of two assumptions. The first assumption is that a customer—or at least that a sufficient percentage of customers in the universe of prospects—conforms to a defined psychographic profile. The second assumption is that the members of a psychographic group will behave as predicted.

Behavior segmentation abandons the process of making assumptions about customers and focuses directly on what they do rather than what they are. Farther along the evolutionary segmentation scale, behavior segmentation creates the market segment directly from the customer's behavior. While the earliest examples of behavior segmentation considered only the broadest measures,

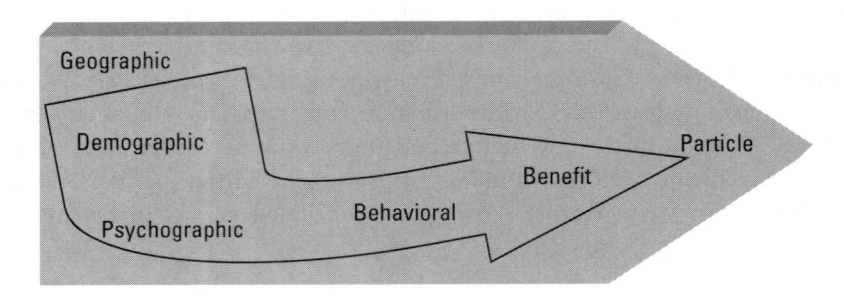

Figure 9.1 Evolution in market segmentation strategies.

such as high-versus-low usage, or per-minute pricing dependent upon the time a call was made, or grouping customers based on international calling patterns, the process contained the seeds of a customer-centric strategy. In fact, when the services offered are indistinguishable between service providers, a situation that constantly threatens the telecommunications services market, usage volume and other patterns are among the only viable behavioral segmentation opportunities.

Predicting customer desires from their ethnicity, marital status, or social attitudes does not yield much value in this market. By studying traits measurable from customer activities, the service provider can identify the profitable high-usage or loyal customer already using the service. But the traditional forms of behavioral segmentation fall short of predicting the behavior of a prospect for whom there is no historical data. To reduce churn, the service provider can offer high-value existing customers bonus minutes or lower prices. To attract new customers who will produce the same value, the service provider can offer bonuses contingent upon high volume or loyalty. Such programs will not appeal to churners and customers who value switching per se, performing a worthwhile screen of prospects for the service provider.

Loyalty segmentation is one example of behavioral segmentation. This form of targeting categorizes customers not by the level of usage, but as the percentage of their overall usage of similar or substitutable services. Consumers can fall into four groups depending on their loyalty status [1]. In declining order, they are hard-core and soft-core, those with shifting loyalties, and switchers. Thus, an Internet café might consider a customer very loyal if most of the customer's visits to any Internet café at all are to that one in particular, and a customer who does not demonstrate loyalty would demonstrate no preference for any café. Except for a very small percentage of the market, most telecommunications customers, especially fixed-line customers, cannot readily choose different service providers capriciously. Though local number portability and promotional programs enable customers to minimize switching costs, there are few incentives for customers to maintain two simultaneous subscription-based solutions to meet wireless or landline access needs. So customers are serially monogamous rather than polygamous, divorcing one service provider in favor of the charms of another.

Another challenge in behavioral segmentation is that behavior changes over time. Establishing customer identities in terms of sales or product mix, for example, fails to capture changes that might have meaning outside of the segmentation process. The business customer whose long-distance revenue drops by several percentage points might demonstrate a change in business methods or policies, or it might simply reflect a rate reduction in the most traveled routes. Changes in customer behavior might reflect organizational changes, such as a reassignment of responsibility for telecommunications purchases. Customers who change their product mix might create behavioral illusions that can affect segmenting. The fixed-line service provider who does not offer wireless services whose market share is growing in a segment under study might be losing customers to wireless only at a slower pace than its competitors are. Suppose the competitors offer both fixed-line and wireless services. If the competitors are successfully attracting the service provider's customers to wireless, and at the same time migrating their own customers, their fixed-line market share will shrink, but their overall market share, and likely their revenues per subscriber, will grow.

Benefit segmentation is the subset of behavioral segmentation that groups customers by the benefits they receive from the services they purchase. This customer-centric classification takes place mainly inside the customer's mind. One temptation of benefit segmentation is to confuse attributes with benefits. Low price is a benefit to customers seeking low price, and unneeded functionality is a disadvantage. For a status-seeker, the reverse is true. Customers buy benefits, and the savvy service provider recognizes that selling benefits rather than attributes means that proprietary features do not create competitive edge unless they also provide unique benefits.

Market segmentation strategies, based on the relatively sophisticated metric of benefits, were responsible for most of the differentiated and curiously skewed pricing that kept the regulated monopoly industry fully funded for decades. Businesses were willing to pay a premium over cost because they gained the benefit of access from consumers with affordable service. Consumers and businesses were happy to pay very high prices for long-distance calls because their only alternative was expensive and time-consuming travel. These pricing irregularities only work in a tightly controlled monopoly market, but they do demonstrate that segmentation decisions can be effective even without competition.

Although marketers lavish most of their segmentation energy on consumer behavior, business buyers are candidates for segmentation as well. In the aggregate, the service provider can still generalize that large-business customers maintain different usage patterns, buy under slightly different value criteria, and expect a different customer interface than do consumers. Demographics and psychographics do affect the individual business buyer's decisions, leading to a swarm of blue-suited computer salesmen decades ago, direct sales representatives who participate in "casual Friday" activities with their clients now, and the perennially enticing steak-and-martini meal. Business account buyers also put their personal beliefs aside to act on behalf of the enterprise and users consume services in the aggregate by account, creating opportunities for geographic and demographic segmentation by industry, benefits segmentation (lowest price, high customer care, customized services) and behavioral (volume, occasions of usage, type of usage) characteristics. Segmentation to reach vertical markets is arguably the business equivalent to profiling lifestyle factors as well. A statistical analysis of customer profiles and behavior from corporate databases can extract customer data for medium and large-business markets or vertical markets and analyze it separately from the data relating to individual buyers.

Once the service provider has identified segments of interest, the variables of accessibility, actionability, measurability, and substantiality help to rank them in importance. Accessibility evaluates whether the service provider can reach and serve customers in the targeted segment. If not, the service provider can choose to focus elsewhere or put an infrastructure in place to meet the segment's needs. Actionability refers to a realistic assessment of whether the service provider has the capability to meet the segment's needs. Measurability assesses whether the service provider can measure the size and other attributes of the target segment effectively. Last, substantiality addresses whether the segment will generate enough business and profits to be worth targeting.

Finally, particle marketing takes segmentation to the individual level, the complete customization of services to the unique desires of the customer. Mining the customer database can offer the service provider opportunities to cross-sell and up-sell to customers based on their own behavior. Furthermore, the analysis of the customer data in the aggregate can create a profile of the type of customer in which to invest marketing resources. The abundance of customer data and the sophistication of analytical techniques raise some stakeholder issues. First, is there a limit to bothering existing customers with new opportunities to spend? Generally, in debates about unwanted marketing intrusions like junk mail, e-mail spam, and outbound telemarketing, the vendor with which a customer has an active business relationship enjoys special status. A service provider can send promotional offers in the bill, and most customers do not consider it in the same category as bulk mail. On the other hand, there is a point at which the customer might believe that finely targeted offers are unwanted enough to seek a new service provider.

The second issue is that the incremental cost of analyzing data already resident in corporate information systems is so negligible that a round-the-clock effort might unearth some unknown marketing opportunity at a very low cost. This could lead to a nearly intrusive analysis of an individual customer's behavior. What is the point at which excellent marketing crosses the line to become the personal stalking of customers?

Lastly, particle marketing, like other information systems support, will become the threshold for marketing, not a differentiator but the price of entry into the marketplace. The information owned by the service provider currently serving the customer is immeasurably more valuable for predicting an individual's inclinations than any data available from any market researcher or outside source. Thus, the sitting service provider should have much better tools than any competitor does to retain the customer's business through perfectly tailored new offerings. If so, then all service providers will bear a customer retention advantage, due to their large and useful stores of competitive information. In theory, then, predictive models for particle marketing should suppress average churn. Or, perhaps it implies that telecommunications services customers are so unsatisfied already that double-digit churn is already lower than it might be without these tools protecting the sitting service provider.

9.2 Statistical Segmentation Using Cluster Analysis

Statistical techniques such as cluster analysis enable service providers to create unique and proprietary market segments from its own customer base for targeting prospects beyond its current customers. Cluster analysis is useful for segmenting markets, because it can establish like-minded groups, or clusters, or segments, within data by looking at the behavior of customers with definable external characteristics. A service provider might assume, for example, that university students are active SMS users, or that prepaid customers tend to use their service outside of the business day, or that the calls originated from airports tend to be long distance. These examples are not real findings from any analysis, just intuition. Therefore, they are much less useful than findings from a genuine database analysis. For one thing, they might not be valid assumptions. Secondly, intuition is available to any competitor and therefore does not provide competitive edge. Third, the benefit of statistical techniques is that they can often produce unexpected relationships that can facilitate effective segmentation and targeting.

Cluster analysis can take two forms, hierarchical and nonhierarchical partitioning. Hierarchical partitioning begins with individuals, and then merges individuals together into clusters. Nonhierarchical partitioning is more complex and works well with large databases. This partitioning technique begins with possible segmentation variables, such as geographic, demographic, benefit or behavioral attributes, and assigns individuals iteratively into groups, reassigning individuals when it improves the clusters. In both cases, the analysis undergoes a procedure to cluster individuals into groups so that their behavior is most similar to the members of the group and at the same time their behavior is most different from members of other groups.

The three phases of cluster analysis are establishing the clusters, interpreting the clusters, and profiling the clusters. Cluster analysis does not develop statistical significance measures, and the technique is often only a first step toward more sophisticated methods. But it produces a sound foundation for further analysis. Cluster analysis and other techniques, including tree-based statistics such as Chi-Square Automated Interaction Detection (CHAID) analysis and Classification and Regression Tree (CART) artificial neural networks, fuzzy and overlapping clustering, and regression techniques, are now the preferred means of selecting target markets. The importance of the newer techniques is that they focus less on what customers are and more on what they do, including their patterns of purchasing behavior, and the value they seek from their purchases.

9.3 Market Coverage Strategies

Service providers can target their segments with their choice of three coverage strategies. Undifferentiated marketing implies that the service provider presents the portfolio in the same manner to all market segments. Differentiated marketing uses market segments to vary the service portfolio and the marketing approach. Concentrated marketing focuses on a single segment and tailors the portfolio and approach to the marketplace to the needs of only those customers.

Under a monopoly, when customers have no choice of service providers, it is possible to be successful using a single or undifferentiated marketing approach. When rates are regulated, a monopolist prefers not to draw attention to diverse, expensive marketing campaigns anyway. An undifferentiated coverage strategy utilizes the large fixed-asset base efficiently, averaging costs over a large base and keeping prices lower than they would be if the portfolio were diverse. Undifferentiated strategies work best when customer needs are homogeneous. But in a highly competitive market, there are few segments whose needs are met with an undifferentiated approach. For one thing, deregulation in the telecommunications market unleashed a diversity of previously unrecognized and unmet customer needs. Furthermore, undifferentiated marketing leads to relentless price competition, a difficult market positioning strategy to sustain.

For example, wholesale bandwidth is approaching commodity status, with its own brokers and clearinghouses. The glut of network bandwidth in the last decade—caused through overinvestment and advances in transmission technologies—created an intensely competitive marketplace that can benefit from the structure imposed by brokers. Companies such as Arbinet, Band-X, and RateXchange participate in wholesale markets in which timing and price are the significant differentiating factors. On the other hand, even in this uncomplicated segment, qualities such as reliability, service quality, contractual terms and financing create a debate as to whether commodity or undifferentiated marketing is possible in any telecommunications market. Moreover, the immaturity of the sector and its lack of standards and infrastructure create an imperfect commodity market.

A differentiated strategy addresses the needs of diverse market segments with varied service packages. Most facilities-based telecommunications service providers in hotly contested markets offer a head-spinning combination of services, price packages, billing plans, and features. The traditional marketing challenge was to create awareness of the vendor and the product; the new marketing challenge is to steer the customer quickly to the very targeted service package that meets the customer's needs better than a competitor's very targeted service package. In this sense, the service provider competes against itself. How many customers leave a first-choice wireless service provider's retail store or Web site to buy a virtually identical package from a competitor, simply because the service provider failed to highlight the package matching the user's needs from a blinding array of alternatives?

Differentiation is not limited to sales or to the product portfolio. Norway's Telenor Mobile uses data-mining technology to sort customers through its customer-care system. The system ensures that particularly valuable customers experience shorter waiting times. This offers first-class service to customers who have earned special treatment. Also, customers who use advanced technologies are assigned the most experienced support representatives. This improves the likelihood of a positive support experience for the customer, and it makes the most cost-effective use of skilled support personnel.

Smaller service providers, especially those without the funding capabilities of industry leaders, might select a concentrated strategy. This strategy combines the benefits of the undifferentiated and differentiated strategies by meeting the unique needs of a small but uniquely targeted segment with a single approach to the marketplace. In conventional product markets, a concentrated approach works well for the very high end of a market, such as consumer goods, automobiles, or highly sophisticated computing technology for scientific or processing-driven applications. In telecommunications markets, concentration is a niche strategy. Correctional communications is one such market, providing specialized inmate payphones and access services. Service providers who choose to operate in this market need to incorporate additional physical security, fraud elimination, limited pricing options, unique information management needs, and local policies regarding usage by inmates. Some of the largest service providers choose not to compete in this market, based on an analysis of its special needs and its potential profitability, paving the way for specialized firms to win contracts against industry leaders.

Candidates for concentration include vertical markets, especially in industries characterized by mid-sized participants. According to a 2002 study by Pyramid Research, the average revenue per subscriber in the small to medium-business segment is 70% higher than in the consumer market. Marketing consultant VIA International maintains that marketing to small and medium-sized business enterprises requires customizing the approach to their needs. This normally means knowing something about the industries they serve. These businesses require more attention than do small businesses and consumers, yet marketing initiatives must be cost effective. Vodafone holds seminars in conjunction with their local resellers in a dozen locations in the United Kingdom. The service provider estimated that one campaign announcing a new service targeted to the transportation industry would gain a 2.5% to 5% response.

In consumer markets, small service providers can benefit from establishing a local presence in localities that the industry leaders are unwilling to commit on-site resources. Service providers seeking to hold a leadership position within a telecommunications market can succeed by redefining the size of the overall market. The niche of interest should be unserved, underserved, or served only as a tiny segment of a large carrier's customer base, where it might very well be unprofitable or otherwise not worth the energy to protect.

9.4 Customer Lifetime Value Management

In theory, the metric of customer lifetime value is the present value (in economic terms) of the profit received during the entire relationship between a service provider and a customer. In practice, calculation of this algorithm shortens the time frame on both ends. If the service provider has an existing relationship with the customer, decisions about customer lifetime value need to focus on future value only. Thus, for example, the marketing initiative that adds, for example, $10 per month to loyal long-term monthly customer accounts that are expected to continue on average for 6 months longer, will have a lower priority than the campaign that adds $15 per month to 3-month customers with accounts that are also expected to continue for 6 months longer. Similarly, estimating many years into the future adds uncertainty without incremental value to the analysis. As a rule, customer lifetime value calculations begin at the present time and focus on only a few years into the future.

The goal is twofold: to invest modestly in retention initiatives for current customers whose lifetime value is very high, and to perform statistical analysis to identify the predictors of high customer value for new customers. Once a set of predictors is available, the service provider can target the segment, especially if the predictors include the benefits that these desirable customers are seeking.

Customer lifetime value management seeks to apply an optimal level of marketing resources to markets in which intervention results in maximum profitability. The lifetime value of the customer is a combination of three factors: the revenue stream, the length of the relationship, and a factor to discount revenues to arrive at present value.

Figure 9.2 depicts the activities that might take place during a common life cycle for an individual wireless customer. The baseline of costs represents the cost of providing service, customer care, and back-office administration. Costs spike at certain points in the relationship, such as customer acquisition, termination, and during marketing promotions. Most of these costs are investments in the revenue stream, which will in all likelihood rise at a faster rate than the present value of the costs. It is evident that a measured investment in the customer relationship will reward the service provider, but that investment needs to be in

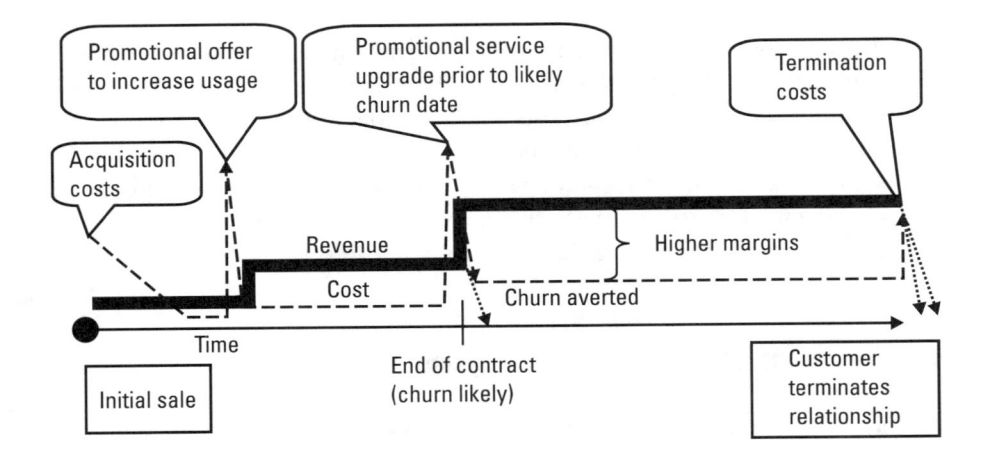

Figure 9.2 Wireless life-cycle management.

proportion to the anticipated average revenue increase from the population targeted.

This chart displays a single customer, slightly oversimplifying costs and revenues to emphasize the concept of lifetime value. In fact, most customer lifetime value analyses average the profiles of an aggregated group of customers instead. The lifetime value for an individual customer is not knowable until the customer has terminated service. Furthermore, this chart shows a best-case scenario; this imaginary customer resisted the temptation to leave when the initial contract was over, and upgraded the service when offered a promotional price. Other customers—who incur the same marketing expenses—do not upgrade their service or switch providers in the face of the incentive to stay. Thus, a proper customer value analysis would include a factor for churn (or the more optimistic label, retention.) This churn number is life-or-death for a single customer relationship, but even for customer segments, reducing churn by only five or ten points changes the lifetime value per customer by a significant amount. According to Jupiter Research, a 3% monthly churn rate for a wireless customer, which is only slightly higher than rates already experienced in the industry, results in a customer lifetime span of 2.8 years. Reducing churn to 2.3% per month increases the customer lifetime to 3.6 years, a 28.6% increase with, roughly, a commensurate increase in customer lifetime value.

The analysis would also target only those customers likeliest to buy with marketing incentives, thus reducing the average cost of marketing. The campaigns can also include prospects with a high likelihood of becoming customers. Customer lifetime value analysis is useful for making cost/benefit decisions about marketing promotions and campaigns. If the average cost of the promotion is higher than the incremental lifetime value averaged over the segment, then the campaign is not worthwhile. This type of analysis reminds marketers that the cost of a marketing initiative covers the entire segment targeted, not just the buyers, but that the revenues that accrue come only from the buyers, diluting the benefits. According to technology industry publisher CNET, wireless acquisition costs per customer reached $22,000 when Deutsche Telekom acquired U.S.-based VoiceStream in 2000. If the U.S. division failed to meet its targets while achieving 32% subscriber growth in 2003, perhaps the assumptions underlying the lifetime value analysis were overoptimistic.

The benefit of the customer lifetime value metric is that it includes all of the elements of interest to service providers: the costs, the revenues, and their duration. Customer lifetime value analysis enables the service provider to identify the customer segments that contribute the most to profit, and identify those segments that represent a net loss to the service provider. The analysis can provide benchmarks over time to ensure that customer segments that were traditionally profitable remain worth targeting. A McKinsey survey determined that wireless service providers who are best-in-class practitioners of customer lifetime management would achieve a 4% or 5% improvement in EBITDA (earnings before interest, taxes, depreciation and amortization) in 2 years [2].

The service provider can measure the lifetime value to date of customers by reviewing only their behavior profiles. A subsequent ranking can group them into quintiles, groups each representing 20% of the customer base. This ranking demonstrates the lifetime value of the top and bottom 20%, or the top 20% and bottom 80%, or five groups of 20%, if desired. One of the characteristics that such an analysis demonstrates is the gradient of decline in profitability between two adjacent groups. If the top group's profitability is decidedly higher than the next group, or if there is a strong decline farther down the curve, then further analysis can demonstrate those characteristics beyond profitability that characterize the top customers and develop service plans to promote to this segment alone.

A wireless subsidiary of AT&T used customer lifetime management analysis to improve the lifetime value of its customers in the face of 25% churn and a slowdown in wireless growth rates. By identifying a higher-than-average quality customer, the service provider was able to reduce churn and create a plan to increase revenues.

British Telecom used cluster analysis, neural networks, association rules, and decision trees to develop and segment small-business customers for a particular service package. Data mining, modeling, and an assessment of customer lifetime value enabled the service provider to double its direct marketing response rate. A customer lifetime value analysis identified one small-business customer segment with high profitability and another segment of customers who required a disproportionate amount of attention but bought services at a below-average rate.

9.5 Making the Most of Undesirable Markets

Telecommunications service providers have only a limited ability to limit their service footprint to lucrative markets. The more realistic option is to identify and target less desirable market segments with a portfolio that will be attractive to those customers and profitable to the service provider.

Customer lifetime management techniques identify both the profitable and the undesirable customer, even customers that produce losses. The methods to attract the best customers work in reverse to discourage the unprofitable prospects. First, a cluster analysis to develop a behavior-based segment should construct a profile of switchers, which the service provider can apply as a prospect-screening device to disqualify these leads through legal and ethical price discrimination and by eliminating front-loaded promotional costs in those markets. One way to discourage switchers is simply to avoid front-end bonuses and promotions in favor of long-term loyalty programs, where benefits accrue slowly and increase in value during a long customer relationship. Switchers generally seek promotions without commitment. Second, the service provider can capture an uncollectible revenues profile. The subsequent customer lifetime value analysis can present a template for screening leads, through pricing or by adding other terms that discourage those prospects who are likely to become nonpaying customers. Third, barely profitable customers can become more profitable if the service provider minimizes the cost of acquiring or serving those customers. The service provider can limit signup promotions, require on-line billing, require payment by credit card or charge interest for late payments on a company-sponsored financial plan, or require prepayment, for example. In few circumstances, the service provider can deny service to customers, such as those who have defaulted on past contracts, if permitted by law.

References

[1]   Kotler, Philip, *Marketing Management: Analysis, Planning, Implementation, and Control*, 7th ed., Englewood Cliffs, NJ: Prentice Hall, 1991.

[2]   Braff, Adam, William J. Passmore, and Michael Simpson, "Going the Distance with Telecom Customers," *The McKinsey Quarterly*, No. 4, 2003.

10

Pricing

10.1 Avoiding Commodity Pricing

Service providers often express concern that telecommunications might become a commodity, that is, that customers perceive the services of all competitors to be undifferentiated from each other. The concern is justified; competition in commodities markets is price driven. Marketers are especially concerned; the way to win price wars is to cut costs, not to invest in superior marketing. Furthermore, price wars produce competitive environments in which market shares remain relatively constant and profits simply shrink, creating customer centered, but unfavorable, marketplaces.

Few telecommunications services act as absolute commodities, but too many service providers have been frustrated by the importance of price as a buying factor. The customer-centered service provider will recognize that the customer's focus on price is less an essential buying criterion and more a demonstration of the inability of service providers to make real perceptual distinctions between services. Trying to launch a low-price strategy is complicated by the downward trend in prices of the past decade in competitive telecommunications markets. According to consultant Booz Allen Hamilton, the average price per wireless minute fell from $0.56 to $0.11 between 1995 and 2002.

The proof that telecommunications need not be a commodity is that some service providers can differentiate their services while others appear to be roughly equal. Some service providers have used the so-called network effect to create switching costs or increase a new market. The network effect occurs when the value of a network-based service is higher as more customers enroll. AOL was so far ahead of its competitors Microsoft and Yahoo! in its instant messaging capability that regulators reviewing its planned merger with Time

Warner precluded the Internet service provider from upgrading the feature until its competitors caught up. AOL's large service base created a universe of potential users who could reach each other, but not the users of competing products offered by its largest competitors. AOL was willing to interconnect and share messages with small ISPs. For a new user choosing between AOL and another Internet service provider, or for someone considering switching to another ISP, the loss of instant messaging capability would be significant.

Another form of network effect took place when Nextel introduced its push-to-talk technology. Business users began to create voice extranets with their contacts. Vendors who used push-to-talk to correspond frequently with their customers were discouraged from switching wireless carriers. The Internet itself is probably the best example of the network effect. As users began to get on-line, it became worthwhile for vendors and suppliers of information to invest in creating pages for the network of users to view. The explosion of new pages made it worthwhile for additional users to go on-line, and so on. Telecommunications service providers who innovate services that generate value as more users sign up will differentiate their otherwise identical services in the minds of customers.

Offering innovative or proprietary content or services is emerging as a differentiator for wireless service providers. In a market characterized by low-end, price-conscious users, Slovakian wireless provider EuroTel increased average revenue per user by an annualized rate of 5% by introducing a combination of customer alternatives, including a service that enables customers to perform bank transactions from their wireless phones.

Furthermore, the service provider can create a brand that capitalizes on the commodity status of a service. A low-end brand could attract the low-price seekers into a segment targeted for its commodity identity. Priced at 5% or 10% lower than the competing services of competitors, a low-end service might appeal to customers seeking the lowest prices. For example, a wireless service provider with better-than-average cost control could offer a service bundle that does not include long-distance, or the enhanced services that are normally bundled, or includes them at noncompetitive rates or in the form of prepaid accounts, uses credit-card payment to eliminate uncollectibles, and requires account management on the Internet. This low-end account could include a price guarantee against the equivalent plans of competing service providers. The wireless service provider could require that the customer agree to receive a designated number of SMS advertisements each month. EuroTel also offers an SMS option that gives the customer a small account credit each time a promotional SMS message is received. A service provider with longstanding or exclusive relationships with advertisers could retain a significant competitive advantage.

10.2 Pricing Strategies

Pricing remains a singularly critical factor in service provider success. It affects the acceptance rate and the profitability of new services, especially for technological innovation. One proven pricing strategy for technological innovation is called skimming, in which the service provider sets a price well over cost to achieve fewer but very profitable sales. As the market develops or as technological upgrades create some obsolescence for the service, the service provider can then drop the price to a smaller margin.

A skimming strategy works best when services have high variable costs. For most asset-intensive telecommunications services, achieving a critical mass of customers quickly is preferable to gaining the small customer base willing to pay for a skimming strategy. Creating a mass market is the best road to the high profits that accrue from a large customer base. Furthermore, even the most disruptive technologies often maintain functional substitutes if not technological ones. Beyond the early-adopting innovators, customers would overlook the new technologies in favor of older technologies with track records and affordable prices. For example, VoIP telecommunications, a completely disruptive technology, is enjoying a slow introduction into the marketplace as a substitute for older networks. Service providers who trumpeted VoIP at a premium price would find most of their target market remaining with older voice and broadband alternatives. Thus, the telecommunications service introduction challenge might be overcome more effectively by low prices, small margins, and creating demand among large numbers of customers quickly.

An opposite market-entry strategy is market penetration, setting low prices for very small margins in an attempt to gain market share quickly. Service providers can also use penetration pricing to capture market share in an existing market. In the telecommunications market, this strategy has several advantages. First, most telecommunications services sectors are very large, so qualified prospects for a service, unless it is highly specialized, are plentiful. Second, sufficient market share often creates a network effect and consequently higher market share, especially if proprietary technology prevents a service provider's customers from interfacing with customers of competitors. Third, telecommunications costs tend to drop over time, so the small margins can grow as long as competitors are not yet in the market or are unwilling to reduce competing prices as their own costs drop. Fourth, telecommunications demand is very price elastic, so usage and revenues generally rise as prices fall. Lastly, more market penetration means better asset utilization when large fixed costs are involved. A penetration strategy is also effective throughout the life cycle of the service, as long as the service provider can sustain adequate margins and differentiate the brand.

Service providers can choose to set their prices at the same level as competitors, a strategy known as competitive or market-led pricing. Service providers who are neither industry leaders nor price leaders have no choice but to follow the pricing activity of those who do exercise leadership in the industry. Assuming that this service provider has a strong cost-leading operational strategy, the market approach can differentiate the services through other elements of the marketing mix.

Customer expectations of price structure and level play a large role in pricing services, especially markets characterized by market-led pricing. More significantly, future price structures will depend on customer expectations from the past, limiting the service provider's pricing options. Will service providers be able to charge by the megabyte for wireless or fixed-line Internet access or e-mail after free and unlimited services have taught customers that surfing is free? All-you-can-use service matches consumption with cost only when the facilities serving the customer bear fixed costs. Under new network topologies, the fixed-price algorithm might become more problematic for providing sufficient revenue against variable costs. The convergence between wireline and wireless access in the United States will create a conflict between diverse but habituated price structures. American customers have been accustomed to measured service for wireless access and unlimited fixed-line connections.

Though service providers appear to wish that new technologies would enable them to restructure prices to usage-sensitive frameworks, marketplace experience demonstrates that customers flock to markets where prices are fixed, and markets with usage-sensitive pricing will stagnate. Fixed versus usage-based pricing is largely responsible for the growth of the Internet sector in the United States compared to countries with usage-sensitive local access, and unlimited service packages are the catalyst that created the critical mass of Internet users even in the United States. Usage-based pricing is palatable to customers only when the prices are so low that customers cannot consume enough service to exceed the price they would pay for unlimited use. The downward trend in telecommunications pricing makes that scenario a likely possibility. Even so, service providers benefit from flat rate and especially unlimited pricing nearly as much as customers do. Unlimited pricing simplifies measurement and billing, though the service provider will need to keep records for law enforcement and data-mining purposes. Hutchison Telecom offered several capped and nearly unlimited plans to encourage customers to use the features of its 3G network, in a market where no competing 3G services were offered.

Usage or packet-based pricing creates the same discomfort among customers as antiquated regulatory pricing models for long-distance calls: customers simply did not know in advance or during the call itself what it would ultimately cost. A Booz Allen Hamilton survey of 700 consumers in five European markets in 2002 found that event-based pricing is much more attractive to consumers than usage-based pricing [1]. Sixty-nine percent of survey respondents would be willing to pay a price per download, such as the download of a song, while only 14% would be willing to pay per megabyte.

Service providers can focus on keeping the benchmark prices low while earning their profits on other services. According to industry observer CNET, Sprint began to charge a flat fee for customer service calls concerning a payment program it offered for subscribers who represented credit risks. The service provider also extended its peak-time pricing for an hour, and began to deduct wireless minutes when customers used more than 30 seconds of airtime to check their balances. AT&T Wireless and Verizon Wireless raised their prices for directory assistance in the midst of a fierce price competition for market share.

Alternatively, service providers can offer free service to gain customer share, in the expectation that other revenue sources will fund their access. In the United Kingdom and much of the rest of Europe, measured local service was the norm when Internet access became popular in the United States. Customers were reluctant to subscribe to ISPs in the United Kingdom until Freeserve offered Internet access at no charge, forcing American ISPs who were seeking to expand their markets overseas to offer free service as well. Anticipated revenue sources included advertiser revenues and sales commissions. Dozens of ISPs, including free, pay-as-you-surf, and subscription services now compete in the U.K. market by offering bonus points to buy products on-line, maintaining loyalty programs, and deploying other customer retention strategies.

DSL providers, whose high fixed costs require significant penetration to be profitable, have discovered that their initial pricing was too high to attract an adequate customer base. Furthermore, DSL is capable of speeds that exceed the needs of most consumers. In Scandinavia, one pricing strategy enabled DSL customers to operate at a fast but temporary rate for the occasional file transfer and other sporadic needs. For about a euro, customers could upgrade their speed, conduct their business, and return to a lower rate. This strategy gives control to customers, something that they desire, enabling them to control both their service level and their costs.

Adoption of DSL access in the United States began slowly and picked up when service providers dropped their prices. Some service providers such as Qwest, BellSouth, and SBC offered several DSL speeds, so that customers could match the speed they used to the price that they wanted to pay. Besides expanding the market for broadband services, the price reductions enabled the service providers to compete more successfully against cable service providers offering broadband Internet access.

Consumers react significantly to price reductions in broadband markets because customers perceive broadband services to be substitutable and undifferentiated. The Yankee Group Technologically Advanced Survey in 2002 found that while one-sixth of broadband households were likely or very likely to switch local service providers for the same price, the number of likely switchers jumped to 60% for a 10% to 15% discount.

10.3 Promotional Pricing

Telecommunications service providers use both push and pull promotional strategies to create demand for their product. A push strategy uses the distribution channels to create demand in the marketplace. Distribution of wireless service through electronics retailers is one way to share the cost of advertising and channel development while creating demand for wireless services. A pull strategy bypasses the intermediaries to create demand, though it does not necessarily bypass intermediary channels to close the sale. Advertising in the mass media, Web site development, and glossy service introductions serve to create demand in the marketplace and "pull" customers to the service.

U.S. mailboxes that were once filled with large checks redeemable by switching to a long-distance carrier are now empty of such enticements. Long-distance service providers recognized that large customer-acquisition costs are not compatible with no-commitment customer relationships, squeezed margins, and falling prices. Service providers have learned that signup promotions must bring in loyal customers, even if that loyalty is coerced. In 2004, Verizon launched a promotion called IN-Network that offered free mobile minutes to its customers whenever they called another customer on the Verizon wireless network. This update of MCI's "Friends and Family" promotion of a decade earlier is noteworthy for several important differences. First, Verizon was already the industry leader. Its ownership of one-quarter of the wireless market would thus result in about a quarter of minutes taken out of allowance formulas, as opposed to MCI's then 10% of the market. It is noteworthy that long-distance service was still nearly a monopoly when MCI launched its program, and the wireless market was much more divided when Verizon offered its version. Though Verizon was leading the market at the time of the launch, 75% of the market was still theirs to capture from its competitors. Second, the promotion applied only to customers whose subscription level was at 400 minutes per month or higher and a one-year commitment. Existing customers—who are often exasperated by promotions directed to new customers only—were eligible for the promotion if they renewed their contracts for a year. Third, the initial promotion was limited to about 3 months. If unintended consequences made it unattractive for Verizon to continue, the service provider could keep it temporary, though it risked the backlash of customer expectations. Also, by establishing a new threshold of pricing, Verizon dared its competitors to offer a similar package. For smaller wireless providers, the impact on in-network minutes consumed would be less of a hardship.

Verizon had undoubtedly tested the concept with data-mining techniques against its customer database and learned that its average customers routinely used significantly fewer minutes than their plans allowed, no matter how many minutes they consumed. Similarly, its tests must have determined that calling patterns would not change so significantly that Verizon's cost structure would change, or that its revenue stream would grow enough to cover any additional costs. Also, its in-network plan enabled it to compete very effectively with Nextel's push-to-talk, offering essentially the same functionality, a wider universe of possible push-to-talk partners, and eliminating the need for its customers to upgrade their handsets to expensive push-to-talk units. Nextel's push-to-talk, which enabled Nextel customers to talk at no charge to other similarly equipped Nextel customers, was targeted to business customers, which made up more than 90% of Nextel's subscribers, according to the company. Business customers are usually willing to invest in high technology, feature-rich equipment, especially when it offers the return on investment likely from push-to-talk. Consumers were a less attractive market, as was demonstrated by the lukewarm reception that they gave similar services offered by Sprint and Verizon. Most likely, Verizon's analysis of free minutes within its network included an estimate of additional revenues from customers willing to upgrade from their own plans to the lowest priced plan that included the free-minutes feature, and a formula eliminating the churn avoided when existing customers made new annual commitments to gain the benefits of the promotion.

10.4 Bundling

Bundling services, that is, offering an array of services to the same customer, has been a vision of the fixed-line telecommunications services industry for decades. Typically, bundling refers to a service package that includes local access and long-distance services, some enhanced services such as voice mail, call forwarding, and call waiting, some bonus usage minutes, and a single bill. Adding Internet service and information services to the package was considered pioneering. Then competition heated up in the wireless market, and competitive local exchange carriers entered the fixed-line market, and customers found most or all of those services in nearly every service package offered by the new entrants.

Such an unoriginal vision of a service bundle pervading the incumbent provider mindset could have been a backlash to the regulations that kept incumbent service providers out of the long-distance market and prevented long-distance service providers from acquiring their way into the local service market. Bundling local and long-distance services did not represent innovation; it recreated the industry of 1960. That the imagined bundled package demonstrated such a lack of imagination in the present competitive environment demonstrates again that the customer drives the threshold of performance expectation, not the service provider.

Defying stereotypes that consumers seek bundles for the discounts they imply, survey after survey discovers that price is not the most significant factor enticing customers to bundle services. A J.D. Power and Associates survey in 2000 found that convenience, defined as the desire to receive a single bill and deal with only one company, was the most important factor leading consumers to desire service bundles. Price was the second factor. Similarly, IDC found in 2003 that 49% of U.S. households would choose a package that offered a single bill, as opposed to 34.2% that would choose the package that offered cost savings alone. Furthermore, a Yankee Group survey corroborated that finding with its own conclusion that 71% of consumers sought a single bill, 60% wanted simplicity, and only 44% were motivated by discounts.

Bundling offers other nonmonetary benefits for both the service provider and the customer. Customers using service bundles are less likely to churn service providers. Customers can select from an array of services, finding the permutation that meets their individual needs without a considerable investment in competitive research.

Because price reduction is not the primary motivator for bundling, risk-taking service providers could provide access to certain services, such as service introductions, on a bundle-only basis, including services with additional fees. Examples include telemetry, proprietary downloads of music or other media, or home-security functions accessible from a mobile phone or from a remote landline.

Local service providers began to offer viable service bundles once long-distance service restrictions were removed. The opportunity was timely for them. Wireless displacement and the popularity of cable broadband service were slowing their access line growth. In the United States, Verizon's Freedom service and Qwest's Choice programs bundled local and long-distance services with the option of adding wireless, broadband, Internet access, and other services. The service providers passed along some of the cost reduction of bundling services, but they wisely promoted simplicity over savings. Both Verizon and Qwest partnered with portal MSN to offer premium Internet capabilities such as instant messaging. Similarly, SBC partnered with Yahoo! in its service bundle, and offered small businesses features such as Web site hosting, e-commerce, and enhanced e-mail management. Vodafone's Live wireless package attracted more than 2 million subscribers in the year following its launch. The bundle's emphasis on gaming, ringtones, and photos positioned Vodafone for success in the 3G broadband wireless marketplace. Vodafone, a second-mover to Japan's DoCoMo and other Japanese wireless service providers, was joined around the world by T-Mobile, Oz, and Sprint in offering packages of proprietary services that attracted customers that want to use their wireless handsets for more than basic voice connections. BellSouth claimed that it improved customer retention by 60% when it offered a bundle that included its fixed-line and Cingular wireless services [2].

Cable companies were also watching customers move toward bundled offerings. In December 2003, Time Warner Cable announced agreements with MCI and Sprint to provide wholesale transmission services in support of Time Warner's full-featured VoIP package. The package included unlimited local and long-distance services, 911 services, directory and operator assistance, and voice mail. The inclusion of digital video entertainment and a reasonable fixed subscription price undoubtedly encouraged more service bundles and sped up the introduction of VoIP technologies in U.S. consumer markets. IP telephony provider Vonage offered cable operators a local telecommunications services package that the cable company could brand for itself. The package included service provisioning and operations, billing, customer care, and even the adapter for the customer's cable modem to connect to Vonage's local network presence.

Telecommunications service providers, having been largely unsuccessful at video-on-demand trials using their own facilities have turned to partnerships to bundle video services within their own packages. Satellite service providers had also seen undistinguished results with broadband Internet access over their networks. Sprint, SBC, and Qwest signed deals with DISH Network and BellSouth, and Verizon signed similar arrangements with DirecTV.

10.5 Price Discrimination

By identifying and segmenting customers into distinct groups, service providers can practice price discrimination, charging different prices to different customers for the same service. Price discrimination takes place because customers exhibit different willingness and ability to pay for identical services. U.S. law prohibits manufacturers from practicing price discrimination, but the law generally does not apply to service providers.

Service providers have historically lowered the average price of services for most customers by practicing price discrimination in monopolistic markets. Competitive markets also benefit from price discrimination, but like all other aspects of marketing, it becomes more complicated because customers control the pricing. The customer-centered service provider will practice price discrimination to satisfy customer needs, maximize revenues, and optimize the portfolio of services in competitive markets.

Price discrimination requires that the service provider have some control over setting prices in at least one market. Because only monopolists can set prices unilaterally, competitors often use behavioral segmentation techniques to discriminate. Price discrimination also requires that the service cannot be resold.

For example, wireless service providers do not have higher costs during the business day than at night or on weekends, but most still charge a higher per-minute rate for calls during peak times. This price discrimination works because it shifts consumer traffic to times of lower network utilization, and those who have the willingness to pay enjoy better network availability any time of day.

Consumers who commit to a high volume of minutes also pay lower rates than occasional users in both the wireless and long-distance markets. Volume discounts arose over a century of commerce because they represented true cost savings for producers and wholesalers, but the cost differences are increasingly irrelevant in telecommunications markets characterized by fixed access costs and network underutilization as technology becomes more robust. Low-usage customers pay more per minute because they expect to do so, and because service providers offer no alternatives. The growing intensity of competition, and lower unlimited usage packages will probably eliminate that vestige of traditional markets over time.

First-degree or "perfect" price discrimination charges each segment at the customer's willingness to pay. This price discrimination largely occurs only within monopolies, because the supplier can successfully control market segments to extract the maximum each customer or each segment is willing to pay. Another example of first-degree price discrimination takes place in one-on-one negotiations between a seller and a buyer, such as used cars, bargaining for large purchases, or on-line auctions. The seller extracts a different price for each unit sold. Though some customers succeed at paying less than they might in other market situations, most customers pay more. The challenge for sellers is to discern what the customer is willing to pay, because the customer has no incentive to be candid with the seller. This form of price discrimination is largely irrelevant in mainstream telecommunications services markets.

Second-degree discrimination includes volume discounts and competitive upgrades. Wireless and long-distance customers pay more per minute for low volumes, and some service providers have enticed new customers with discounts for switching from another service provider. Service bundling creates price discrimination when customers receive discounts for multiple services.

Third-degree price discrimination uses market segments to accomplish premium pricing from those customers willing to pay. Peak-hour pricing and preventing businesses from buying residential services are longtime examples of third-degree discrimination in the telecommunications market. Third-degree price discrimination is possible in a market segment that demonstrates price inelasticity. The business customer making an international call to a client will pay much more than the service provider's cost, assuming that all service providers charge roughly the same price for the call. The same business customer might go home and choose not to make a similar international call to a friend at half the price, selecting a less expensive alternative, such as e-mail, or choosing not to communicate at all.

Price discrimination most benefits those customers who pay the lowest prices, but it also benefits all customers, especially in asset-intensive markets. The low prices bring in more customers who might not have bought services at an average price. Price discrimination based on time-of-day enables the customers to optimize network utilization, creating a proper investment and operations level. Profits from price discrimination might allow service providers to invest more in research and development or provide lifeline services. Price discrimination might increase the overall level of consumption, encouraging service providers to build bigger networks and other infrastructure and gaining economies of scale, lowering prices to all customers. Nonetheless, customers sometimes balk at paying premium prices for equal services. When cable service provider Comcast tried to reward customers that bundled cable entertainment and broadband Internet services with an Internet services discount, the move angered Internet-only customers and generated considerable negative publicity from a relatively tiny stakeholder segment.

10.6 Pricing for Competitive Parity

In most markets, the industry leader is the price leader. Price reductions, increases, or structural changes are a defensive strategy for the leader. When costs drop, the industry leader needs to adjust prices to fend off possible churn if another competitor were to lower prices as an offensive strategy. Once a customer is lost, it is many times more difficult to win the customer back than to have kept the customer from leaving in the first place.

In the wireless market, price leadership has shifted from one service provider to another, and market shares adjusted as a consequence. AT&T Wireless made up for its late entry to the marketplace with a pricing structure that eliminated long-distance charges and roaming charges. It gained market share immediately and became a major player, retaining its leadership until other service providers were able to establish the infrastructure they needed to meet AT&T's new threshold. U.K. wireless service provider Orange entered the market late, unknown, and with technology similar to a service provider already in the market. Along with a large network investment, Orange priced very aggressively, offering about a 30% reduction in price compared to its competitors [3]. The service provider took about a third of the market growth after its entry, in spite of below-average network coverage. Competitors Vodafone and Cellnet followed Orange's spectacular entry with price reductions of their own.

In the long-distance market, price leadership as an offensive strategy worked well for Sprint in chipping away at AT&T's industry-leading market share. Sprint lowered prices, certainly, but its major contribution to the industry was to simplify prices, charging for long distance for the first time on a per-minute basis. Once a price follower, full-service provider Verizon began restructuring prices defensively once its leadership position made it vulnerable to attacks by competitors.

Game theory is the study of strategic behavior between rational and interdependent competitors that seek to maximize their own positions. The participants in games of strategy hold imperfect information, but they must choose whether to cooperate or compete. Service providers need to anticipate the pricing moves of competitors before making their own pricing decisions.

Figure 10.1 demonstrates that service providers need to anticipate and react to the pricing initiatives of others. Telecommunications services continue to be extremely price elastic; that is, when prices fall, volumes rise sufficiently to overcome the price reduction. This elasticity means that raising prices is not worthwhile; volumes and ultimately profits would fall. Game theory predicts that there is a desirable strategy in every competitive transaction. For example, if a competitor lowers the price of service, volumes rise at a faster rate than prices fall. Matching the price reduction immediately results in more revenues for both competitors and retains market shares. Failing to match the price reduction results in lost customers for the second competitor, because customers of individual undifferentiated services are very price sensitive. The result is loss of market share, lost revenues, and a loss of profits. Similarly, reducing prices when the competitor has not acted first will move market share away from the competitor,

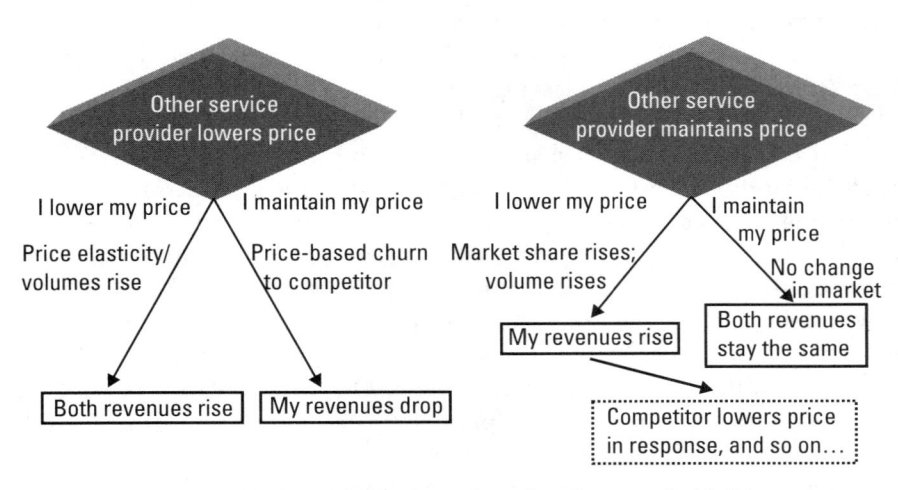

Nash equilibrium: Lowering prices (cooperation) is better than maintaining them no matter what competitor does.

Figure 10.1 Game theory pricing strategy.

adding to both volume and market share. But undoubtedly the migration of market share is temporary, as all games within competitive markets are repeated, changing the decisions of the players.

References

[1] Meier, Helmut, et al., "How Wireless Carriers Will Make Mobile Data Pay," *strategy+business*, eNews, October 23, 2003.

[2] Laughlin, Kirk, "Bundling Branches into Wireless," America's Network, Vol. 106, No. 11, 2002, p. 24.

[3] Gurumurthy, Kalyanaram, and Ragu Gurumurthy, "Market Entry Strategies: Pioneers Versus Late Arrivals," *strategy+business*, Issue 12, Third Quarter 1998, pp. 74–84.

11

Customer Loyalty, Retention, and Churn

11.1 Predicting and Limiting Churn

Any discussion of loyalty in the telecommunications service market begins with the much-discussed metric of disloyalty called churn. Service providers calculate churn by making a ratio between customers leaving during the period under study and total customers during the period. Slightly oversimplifying, when 20 customers leave a service provider that averages 100 customers, churn is 20%.

During periods of very high growth, the appearance of new subscribers mitigates churn. New subscribers produce revenues and profits that compensate for those lost to churn. New subscribers increase the size of the customer base, which is the denominator of the ratio, making churn slightly less striking. Of course, keeping the existing customers while gaining those new ones would be much better. But if churn is destiny, then simultaneous growth makes it less painful, and service providers had about a decade to face churn and learn how to cope with it while maintaining profitability during a period of high growth.

Recently, growth has slowed in many telecommunications markets still affected by high churn rates. Slower growth will not only make existing churn a larger problem, it will probably make churn rates larger. Slow growth means that those service providers who want to gain customers will have to take market share from their competitors, resulting in more intense competition and more incentives for customers to switch service providers. Industry consultant McKinsey & Company estimated in 2003 that 80% of a service provider's future wireless customers would come from churn, not growth. The maturation of the industry also means that new subscribers—the late-majority technology adopters—tend to demonstrate less interest in costly and more profitable new features. The upshot is that wireless service providers will have to fight harder to capture less desirable customers.

Churn is costly. Losing one customer means that a service provider needs to acquire a new customer just to stay even. Consultant AMS estimates that it costs service providers $60 to retain an existing customer annually but it costs $400 to acquire a new one. Furthermore, longstanding customers are more profitable, because they cost less, spend more, and have already passed the test of creditworthiness as compared to new, unknown customers. Strategy Analytics estimates that a 1% increase in churn in the worldwide wireless market translates into $2 billion in subscriber acquisition costs just to maintain the customer base. Consulting firm Bain & Company has claimed that companies can boost revenues up to 85% if they retain only 5% more of their best customers. Churn interrupts the amortized recovery of the acquisition costs invested in the customer. It removes the opportunity a service provider might have taken to upgrade a service contract after an enduring relationship.

Churn takes three forms: unavoidable, involuntary, and voluntary. Unavoidable churn occurs when the customer leaves the service area of the provider, and involuntary churn occurs when the provider terminates service for nonpayment, fraud, or other reasons. By far most churn is voluntary, and it is of paramount interest to service providers.

The intuitive response to the churn problem is that service providers need to improve their customer satisfaction metrics. Customer dissatisfaction is related to churn, but it is not the only measure, nor the best one. Mercer Management Consulting found that 80% of customers of a subscription service described themselves as "very satisfied" or "quite satisfied" within a year before they churned.

Bain & Company, a leader in loyalty research, learned after many years of study that intuition could be improved by analytics. The consultant found that the question "How satisfied are you with Company X's performance?" was not a very good predictor of growth. A better question was "How likely is it that you would recommend Company X to a friend or colleague?" The customer that recommends a service provider is thus a "promoter," and the best companies have 80% more promoters than detractors. For most companies, the promoters outnumber detractors by only 10%.

Due to the intensity of competition and the abundance of major service providers willing to invest in customer acquisition, wireless service and Internet access services are the telecommunications services sectors most affected by churn. Internet service providers do not publicize their churn rates, but industry observers estimate annual churn for ISPs at more than 50%. Wireless churn in the United States and Europe is estimated at 2% to 3% per month, or about one-quarter to one-third of the customer base leaving their service providers each year. Prepaid wireless churn is higher, about 6% to 8% per month in the

United States. Though prepaid represents only a small percentage of U.S. wireless customers, the high rate of churn adds about one-half of a percentage point to overall churn levels.

Market researchers have predicted that number portability in fixed-line and wireless markets tends to increase churn. Service contracts hold penalties for early termination, so they certainly had an opportunity to retain most customers throughout the duration of the contract. That 1-year commitment in a market without local number portability enabled service providers to get customers attached to their numbers, which probably kept some customers that might otherwise have churned for additional years. Without the protection of the need for new phone numbers keeping customers loyal, or at least behaving loyally, churn rates are a greater challenge. Atlantic-ACM predicted that local number portability would increase preportability wireless churn rates of 25% to 35% by another one-quarter to one-half.

Churn rates are lower among service providers focused on business customers. U.S. wireless providers Nextel and United States Cellular kept their churn rates about 15% below industry averages. The lower rate of business-customer churn that these service providers sustained might stem partly from customer loyalty and partly from stringent lead qualification. Nextel maintained high credit standards before signing customers, reducing involuntary churn, and did not support a prepaid program and its associated high churn rates.

Figure 11.1 shows that consumer likelihood to churn rises quickly if churning would offer a discount. Nevertheless, more than one-quarter of customers stated their intention to switch providers even without a discount. For individual wireless service providers, the data diverges. Business provider Nextel demonstrated the lowest potential churn at each price point. T-Mobile, which positioned its service as the low-price alternative, was most vulnerable to churn at each price decrease among eight U.S. providers studied. These findings are intuitive, as the low-price-seeking consumers that supported T-Mobile should be more likely to churn than business customers that supported full-featured Nextel. Gartner Group identified price, service quality, and coverage as the three top reasons for wireless churn. Most price-related churning occurs when customers perceive the services offered by competitors to be identical. Price-driven churning will continue to occur as long as the arms race of competitive parity keeps service portfolios equivalent. Price-driven churning will stop when undifferentiated services are so inexpensive that choosing between service providers is unimportant to customers.

Market researcher Walker Information found in 2003 that though three-quarters of U.S. telecommunications customers were at least "satisfied" with their service providers, only 28% were "truly loyal," meaning that they planned to keep buying service from their current provider. Given that that many customers churn within a year after reporting similar satisfaction, even the loyal

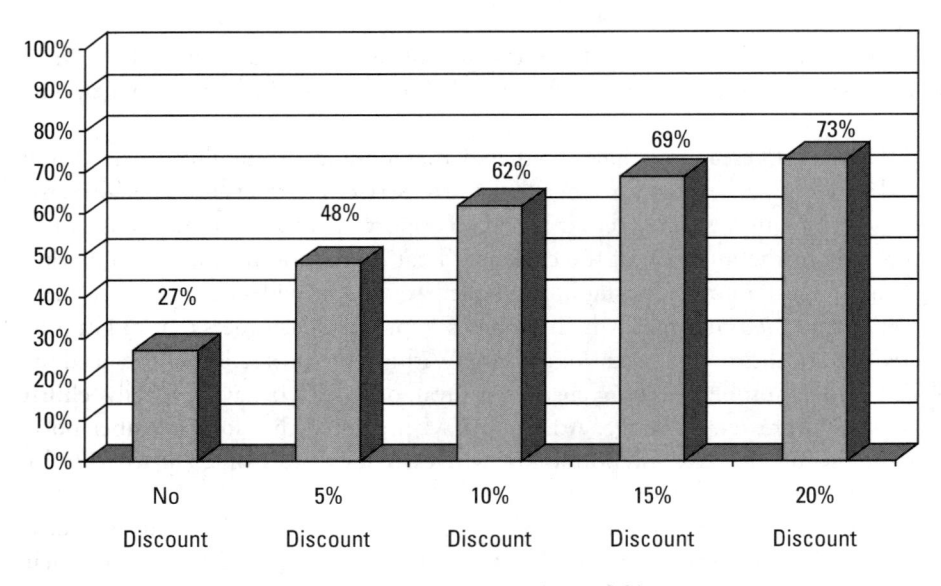

Figure 11.1 Consumer churn for wireless discounts. (*After:* [1].)

customers would be at risk. Walker Information reported that nearly half of the customers surveyed planned to continue the service relationship only until something better came along, and fully one-quarter of the study group were at such high risk that they were ready to begin purchasing from a competitor at any time. Loyalty was similarly dismal across sectors, with only 23% loyal to local service providers (taking into consideration that most of them did not have practical alternatives to their present service provider, a source of dissatisfaction on its own). Long-distance providers earned only 28% loyalty, and Internet service providers and wireless providers both received loyalty ratings of 30%. The study found the primary driver of loyalty to be a customer focus.

11.2 Strategies for Maintaining Loyalty

Loyalty is not the same as retention, though it is related. Retention is a simple ratio, the inverse of churn. The service provider with 100 customers at the beginning of the study period and 80 at the end has a retention rate of 80%. Loyalty occurs when customers remain with a service provider even when other service providers offer equal or better solutions. A customer who stays with the least expensive and highest quality service, or with the only package in the marketplace that perfectly matches the customer's needs, is rational, not loyal. The customer who prefers a service provider's brand to a competitor with an equal or potentially better offer is loyal. Loyal customers are usually desirable, but retention of profitable customers, loyal or not, is necessary for profitability. In fact, neither loyalty nor retention is always advantageous.

Service providers depend on data-mining techniques and their customer-relationship management software to identify likely churners in time to recapture their subscriptions. Technology tools have become necessary, but they are not sufficient to constitute a churn reduction program. Strategies to meet the needs of specific customer segments, vigilant monitoring of competitive portfolios to seek differentiation opportunities, and experimentation with retention techniques are critical management initiatives.

For an individual customer, churn management covers the entire customer life cycle. Before the sale, service providers should target desirable market segments aggressively. Service providers can target those segments that are likeliest to churn using low-cost channels and offering a portfolio that would be profitable no matter how long the relationship lasted. Marketing expenditures such as signing bonuses should remain minimal for likely churners. For example, the service provider can use Internet-only offers to funnel sales through a self-service delivery channel and automate the sales transaction. For all sales, service providers should attempt to optimize acquisition costs to reduce the time it takes to recover the marketing investment. A welcome package that coincides with the first bill can reduce customer anxiety about the payment. If the welcome package offered a short-term upgrade or an incentive to use premium services coupled with a short expiration date, the bill might also serve as a sales call. Cross-selling services to create unique bundles can increase the customer's lifetime value and serve to prevent future churn.

Throughout the service relationship, the service provider can use surveys periodically to test the subscriber's intention to churn, though too much attention to customer satisfaction might drive some customers to a less intrusive supplier. When the customer signals an intention to churn, either because a contract is ending or because usage is down significantly, the service provider can intervene to save the relationship. Behind the scenes, the service provider can monitor the customer's usage to detect changes in patterns. The customer that consistently goes over the allotted minutes is a candidate for a plan with a higher minimum and a lower overall price. If the service provider does not offer a better plan to that customer soon enough, someone else will.

Finally, when the customer contacts the service provider to terminate the relationship, the interaction represents the last opportunity that the service provider gets to change the customer's mind. Few service providers with large customer-service infrastructures allow customers to terminate service on-line or without interacting with a customer service representative. The one area of customer care that need not be easy to complete is service termination. One ISP assigns its best sales representatives to take the calls from customers trying to terminate service. While the customer is still on the telephone, the sales representative reviews the account for opportunities to win the customer back. The sales representative uses the customer's reason for leaving as leverage to demonstrate the importance of staying: "You've lost your job? Did you know that you can find a job by searching the Internet using our service?" or "Your children are wasting time on-line? Did you know that they could use our service to do their homework?"

It is admirable to assign resources to recover the customer at the end of the relationship, but too often the customer notifies the service provider after signing a new subscription. In many telecommunications services markets, the customer does not have to notify the losing service provider at all. Furthermore, exit interviews are useful, but few surveys return data that is unequivocal.

One ISP used database analytics to link customers who were unable to connect to the server successfully with a propensity to churn. Armed with this relationship, the ISP could set alerts to identify customers having trouble connecting, in time to offer the customer rebates, upgrade the network, or do whatever was necessary to eliminate the cause of churn.

Nextel used attentiveness to maintain its customer relationships. The service provider set a goal to answer inbound calls for orders within 30 seconds, maintain an abandonment rate (not answering the call before the customer hangs up) under 5%, and accomplish a turnaround of less than 4 hours between customer activation and fulfillment [2]. A few weeks after service activation, Nextel called the customer to make sure that service was satisfactory. Two months after activation, Nextel sent a welcome kit and a card thanking them for their service. A month later, the customer received a newsletter. The service provider also tied employee bonuses to customer satisfaction.

In consumer markets, bundling appears to reduce churn considerably. SBC has stated that churn dropped by 9% when long distance was part of the access line subscription. When the customer's service package included DSL, churn fell by 61%. A package with both long distance and DSL reduced churn by 73%. Nonetheless, service providers who are thrilled to fend off churn with bundles need to recognize that it is probably a temporary solution. A combination of regulatory relaxation and technological investment has made bundling of access and transmission services a relatively new marketing tool. The reduction in churn from bundling services will be a competitive advantage only as long as a single service provider in each market offers the bundle that a customer wants. Permutations with VoIP access and transport, cable and satellite entertainment, and a variety of fixed and wireless broadband alternatives will level individual markets with competitive parity. Undoubtedly, churn will then rise unless service providers can find yet another path to differentiation.

Service providers can limit churn in prepaid services by adding the features that keep postpaid subscribers from churning. Enhanced services like SMS, roaming and data access are not available on many prepaid plans. A service provider can gain temporary competitive advantage by offering a full-featured prepaid plan. One advantage of additional features is that customers will decrement their accounts faster and recharge more often. Besides enticing prepaid subscribers to remain with the service provider, these features might keep less-reliable payers on a prepaid plan where they can manage their expenditures rather than migrating them to postpaid services where they might create involuntary churn.

According to Australian marketing consultant De Weaver and Partners, loyalty programs can include affinity, appreciation, coalition, partnership, rebate, and rewards. Affinity programs do not use rewards at all. They use areas of mutual interest to attract customers. Affinity programs require very strong brands to attract customers, and are uncommon in the present undifferentiated telecommunications industry. Appreciation programs give customers more of the product as a reward for loyalty. Frequent-flyer programs in the airline industry created a surge in appreciation programs. Rollover minutes and bonus minutes are examples of appreciation. Coalition programs partner with other suppliers to pool customer databases and make targeted offers to selected demographics. Partnership programs work like coalition programs, but customers can choose their rewards from either partner. Rebate programs return money to loyal customers, and rewards programs allow customers to collect prizes unrelated to the service they bought in return for their loyalty.

AT&T Broadband, now a division of Comcast Cable, created a rewards program to diminish churn, reduce service downgrades, and increase service upgrades for targeted customers. The program selected high value, highly vulnerable customers based on their service packages and the competitiveness of the markets in which they were located. The program featured a catalog of merchandise redeemable for points, and was supported by direct mail, call centers, and the Internet. The program reduced churn at a statistically significant level, by 27%, and reduced service downgrades in the targeted group by 44%. Within 3 months, the program had recovered its costs.

11.3 Limiting the Costs of Churn

Service providers can reduce the level, or at least the cost, of involuntary churn by taking proactive measures. Migrating risky customers from postpaid to prepaid services or direct debit accounts ensures that all revenues due are collected on time. Ensuring that customers of postpaid services can check their minutes during the billing period—if only on the Internet or through automated callback to save customer-care expenses—helps to reduce involuntary churn.

Service providers can recapture some customers from unavoidable churn. Some customers wrongly believe that a move will take them out of the service area. Others, unaware of low-usage plans or prepaid service, might want to cancel after watching their minutes expire every month. The service representative at the terminations desk should be able to recover these customers before they are already exploring the offers of competitors. By that time, the chance of winning them back drops exponentially. These recoveries will occur episodically when service representatives are creative, or routinely because of service representative training and intelligent software design.

McKinsey & Company discovered that wireless customers whose monthly bills fluctuated substantially were most likely to churn [3]. Customers whose usage varied by more than 23% made up 20% to 40% of subscribers, according to McKinsey. The consultant suggests that customers with volatile bills might prefer an annual package in which billing is calculated as the average of previous bills, with timely notifications of overages well outside the averages.

McKinsey also suggests that service providers adjust their focus away from only customer defection to migration, that is, customers still subscribed but using the service less. McKinsey argued that a focus on migration would have two to four times the impact of a focus on defection alone. Migration is a leading indicator of subsequent churn, and it is directly tied to economic value.

Churn increases acquisition costs by causing service providers to replace customers more often. Overly aggressive customer acquisition programs can actually add to churn, by neglecting credit checks, by attracting frequent churners, or by signing customers at a faster rate than corporate systems and networks can absorb their arrival. Sales objectives that do not match the marketplace or that drive the sales force to overbook sales can also add to unnecessary churn.

In an adjunct to, and a reversal of, cutting churn, service providers can also seek to limit their acquisition costs by managing them as carefully as any other expense item. The more churn a service provider faces, the more value a reduction in acquisition costs will have, as long as the costs eliminated do not reduce customer lifetime value. An acquisition investment that does not yield a new customer has the same impact on the bottom line as any other waste. Furthermore, the cost of acquiring each new customer is the average of the entire program that brought the customer to the sale. The less successful the program, the higher the average cost. The higher the cost, the lower the lifetime value of the customers acquired. Acquisition programs should be limited to those prospects that are very likely to buy and that would not have purchased the services if the acquisition investment had not been made. An acquisition program is only successful if the incremental customer lifetime value exceeds its fully allocated cost. Even then, investments have opportunity costs. The program that earned a certain return of incremental lifetime value must compare favorably to other programs that might have brought in a higher return.

It seems hardhearted, but marketing investment is wasted on those long-time loyal customers who are very unlikely to churn to a competitor, or who are most likely to give up service in an act of unavoidable churn without switching to a competitor. Data mining can also assist the service provider to identify those existing customers who would be very unlikely to churn no matter what competitive offer might be before them. Incumbent service providers might use analytical tools to develop a profile of potential customers who are unlikely to churn in a competitive environment. Intuitively, for local service providers, that segment might include factors such as decades of service without changing numbers, an account without enhanced services, or usage patterns that indicate a household without Internet access. For wireless providers, one loyal segment might include late adopters who are interested only in the safety of having a wireless phone, not using it. Instead of presenting that segment with generic low-usage plans, the service provider could offer a very inexpensive and basic handset, partner with other technology providers to offer other features that would appeal to the segment, and eliminate confusing features like SMS that these customers shun.

Nonintuitively, analytics might uncover other behavioral characteristics for the profile. For customers unlikely to churn, the service provider can save marketing expenses by overlooking these customers when distributing promotional materials that encourage retention. Of course, it is essential to examine the results of this sort of segmentation not only as a component of overall churn but within the group itself, to ensure that the targeting was directed appropriately.

One underutilized element of customer retention is winback, that is, the recovery of customers lost to dissatisfaction. Though reviving the relationship after the customer has departed is much more difficult than keeping them from leaving, these customers have two characteristics that make them more desirable than unknown prospects. First, the service provider can review the customer's historical data using data-mining techniques, retaining some advantages of an incumbent service provider. Second, the glow of uncertainty for the untested competitor has abated for the customer who has already defected and is probably building up an arsenal of dissatisfaction with the new choice. Excellent competitive differentiation, keen data mining, and dutiful repentance can restore the service provider to the customer relationship. Best practices for winning back departing customers include proactive behavior and commitment. The service provider must act quickly in the hope that the customer will return before committing to a competitor. A tepid form letter does not capture attention or communicate the sincerity of a repetitive, personalized campaign. Some successful winback campaigns empower service representatives with a budget or authority to remove barriers to the customer's return. On the other hand, these campaigns can attract unwanted attention from regulators if the successful service provider is drawing back a large proportion of defectors during a deregulatory initiative. During the introduction of carrier long-distance preselection, former monopolists such as France Télécom, Deutsche Telekom, and Ireland's eircom were criticized for or blocked from continuing their aggressive winback campaigns.

11.4 Benefiting from Churn

It seems counterintuitive that a service provider can actually benefit from churn, but it is possible. Studies of customer lifetime value have demonstrated that few customers produce a large profit, most customers generate minimal profit, and a sizable group of customers actually reduces profitability. Fujitsu Consulting has said that the top 20% of telecommunications customers spend 30 to 50 times more than the lowest 20%, and found, in a review of customer purchasing behavior, that more than half of customers are unprofitable. Loss-producing customers overuse customer service, spend little on the features and services that contribute most to profits, and use the service provider's facilities less efficiently, for example, wireless customers roaming frequently on partner networks. This is one reason that loyalty alone—or even retention—should not be the overriding goal of the relationship. A more important goal is to maximize the retention of desirable customers and migrate or churn customers who subtract from profit.

Most of the data for calculating the lifetime value of a customer, other than the eventual length of the relationship and the reason it will end, is available while the customer is active. Service providers can conduct activity-based analyses to identify their costliest customer-care activities and match it to specific customer records, for instance. Other data, including the predicted lifetime of the relationship, uses segmentation strategies to match the projected lifetime against former customers with the same attributes. To the degree possible, costs such as anticipated bad debt require individual or very targeted analysis rather than averaging over the entire database, because average parameters might make a profitable customer seem unprofitable, and vice versa.

The churn management process evaluates corporate data in the aggregate to develop segmentation strategies and improve business processes that will reduce overall churn. Figure 11.2 depicts the sequence of activities. The process begins with all corporate data, not only customer information. The process matches customer-service complaints against operations data to identify areas of service deficiency. Customer surveys can compare internal satisfaction results with benchmark industry data. Termination surveys can provide insight concerning why customers churned in the past, though too late. Call data can include customer usage, service quality, or call attributes like duration or timing. Customer usage can identify telltale patterns that precede churning, such as usage declines or costly overages. Customer account history can include payment patterns.

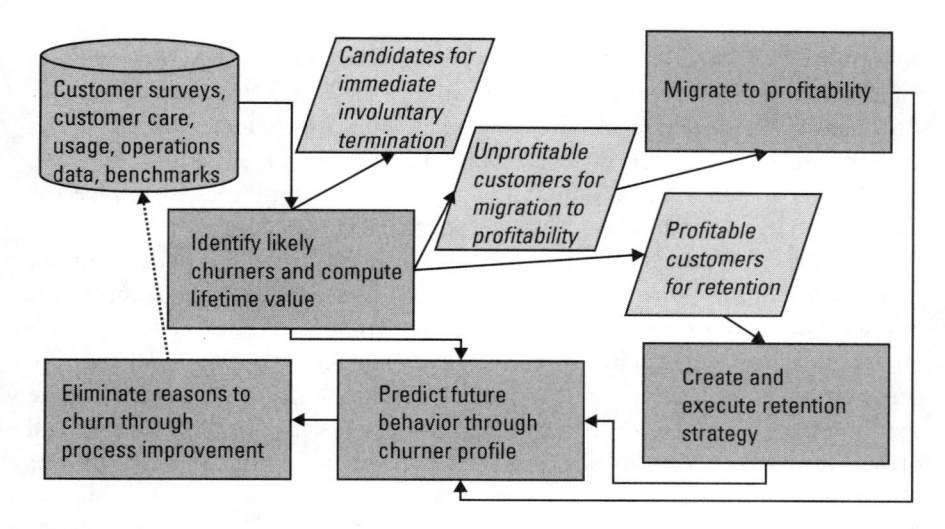

Figure 11.2 The churn management process.

The service provider can isolate likely churners and assign them to segments that match their profitability. Unprofitable customers should receive further examination to determine whether termination is appropriate or whether they might become profitable in a different service arrangement. Profitable customers merit marketing investment to retain them. This investment might include premium customer care, bonuses based on account longevity, and gifts. Discounts constitute an inappropriate investment, unless the discount represents early entry into a new pricing program that would treat new customers differently than the contractual ones. First of all, discounts eat into customer lifetime value. Second, loyal customers do not churn for discounts, by definition. Third, temporary discounts create expectations and could adjust the customer's buying criteria, something the service provider would not want. After all, a match between the service provider's attributes and the customer's present buying criteria is what made the customer loyal in the first place.

All of the knowledge gleaned from the existing customer base can enhance the algorithm to segment customers in the future, and to evaluate the future behavior of present customers when they signal an intention to churn. Furthermore, the process can identify the segment of unprofitable customers to avoid marketing investment or to steer them to packages that would foster profitability.

The process can also identify those controllable service and process elements that cause dissatisfaction among customers who eventually switch providers. To the degree feasible, the service provider needs to invest in weak business processes and service quality to eliminate those causes of churn. Bell Nexxia, a division of Bell Canada, serves the business market with high-speed data and Internet solutions. The service provider surveys the entire customer base annually, less to test elusive satisfaction levels and more to elicit suggestions about how to improve the solutions it provides. Taking the survey results to action, Bell Nexxia includes an accountability program for employees to resolve the issues raised by the customers.

Churn in some sectors might be due to a self-fulfilling prophecy. Prepaid wireless, sustaining churn rates of two to three times the already high churn rate of postpaid, earns little marketing investment from most wireless service providers. Yet Telecom Italia Mobile sustains a large majority of its customer base from prepaid and is the leading wireless carrier in Europe. Industry observers note that prepaid does not achieve the average revenue per user (ARPU) of postpaid plans, but that view is simplistic. First of all, maintaining prepaid accounts does not require the costs of billing and payment processing, handset subsidies to bribe customers to sign annual contracts, or bad debt. The opposite occurs; breakage (leftover minutes in the account) allows the service provider to use the revenue already collected from the prepaid customer before incurring network costs associated with providing the minutes of use. Moreover, most ARPU analyses of prepaid and postpaid usage do not appear to account for differences in how the customer utilizes the network. For example, a prepaid user that cannot roam will not accrue expensive transfer costs to other service providers or the inconvenience of completing inter-company settlements. Also, prepaid customers represent an opportunity for pricing structures on a clean slate. Typically, minutes against prepaid accounts cost more than for postpaid customers. After account activation, few prepaid packages include a large bundle of free minutes or minutes at lower prices. Most prepaid customers are consumers or workers, not managers, so they tend to use the service during hours that the network is underutilized anyway. Prepaid pricing structures enable the service provider to schedule peak and off-peak periods without the risk of annoying the most lucrative segments of the postpaid market. It is as if prepaid is a separate company.

References

[1] TNS Telecom, http://www.tnstelecoms.com.

[2] Marek, Sue, "Attacking Churn with Education," *Wireless Week,* Vol. 8, No. 30, 2002, pp. 22–23.

[3] Bienenstock, Robin, Paola Bonomo, and Richard Hunter, "Keeping Mobile Customers," *The McKinsey Quarterly*, No. 1, 2004.

12

The Customer Relationship Management Process

12.1 Customer Relationship Management Systems

The phrase "customer relationship management" or CRM is sometimes used to describe a software application, at other times it means one-to-one marketing rather than to a segment, and at other times it refers to the general process of managing the various interactions a customer has with a service provider. The overall process was once simply called customer service and more recently called customer care, when the customer became an important focus. The overall CRM process benefits from automated support in the form of customer relationship management systems, and one-to-one marketing or particle marketing describes the techniques used for creating market segments of a single customer. Unless otherwise specified, customer relationship management will henceforth mean the process of managing the customer relationship and customer relationship management systems will mean the automation that supports the process.

According to consultant Booz Allen Hamilton, customer services can consume up to 20% of a service provider's operating expenses. Yet loyalty researcher Walker Information found that:

- Less than half of telecommunications customers believe that their service provider really cares about customers;
- Less than half of customers actually prefer their own service provider to other brands in the sector;
- Only about 4 in 10 customers believe that they get "excellent" or "good" value from what they pay for their telecommunications services.

Perhaps it is the telecommunications service providers that are not getting excellent or good value from what they pay for customer relationship management automation. A Gartner Group survey found that 68% of customers had left a service provider because they were upset with the poor treatment they received. Software support in the form of customer relationship management systems makes up a large portion of the cost of customer care. Consultant Bain & Company found in its cross-industry Management Tools 2003 survey that customer relationship management systems represented the fastest growing by far of the 16 management tools it studied.

CRM systems perform analyses on customer data such as usage and billing information, demographic profiles, payment patterns, purchasing history, and profitability. Vendor applications differ in perspective; some vendors focus on customer retention or profitability, others emphasize intersystem compatibility, and others promote the system's availability to employees and partners through intranets. These systems include tools that conduct sales force automation, call-center support, business intelligence and analytics, product and market management, order management, service life cycle management, and marketing campaign management support. Virtually every aspect of the marketing function can find support in a CRM system, if the requisite data, financial commitment, and cultural predisposition are present.

Documented success of automated CRM systems is much less apparent than the enthusiasm for it, especially in the context of its large investment. Figure 12.1 demonstrates that only one-sixth of customer relationship management systems succeed at producing quantifiable benefits. AMR Research determined that 12% of CRM projects fail to go live, and thus became a technology failure. An additional 47% accomplish successful technology implementations

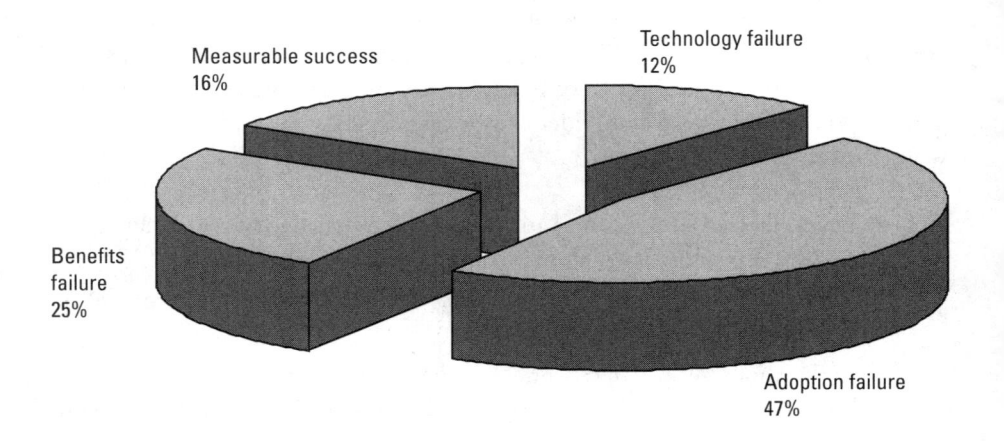

Figure 12.1 The CRM gamble. (*After:* [1].)

but the sponsoring organization does not adopt the system successfully and consequently the systems do not create business change, causing an adoption failure. The remaining CRM projects, constituting one-quarter of the total, are adopted, but fail to demonstrate tangible business benefits, creating a benefits failure.

Aberdeen Group has concluded that the high level of dissatisfaction with automated CRM systems is due in part to mismatched expectations. Though companies might have embarked on CRM implementation to gain competitive advantage and revenue growth, much of the actual benefit reported by users was typically in the areas of cost control and productivity. Aberdeen suggested that such productivity improvements represented a leading benefit necessary on the path to revenue and market share increases. One modestly reported expectation and benefit was defensive: a CRM system, as costly and difficult as it might be, was considered necessary only to achieve competitive parity.

A 2003 Forrester Research survey concluded that three-quarters of surveyed North American executives were very satisfied or somewhat satisfied with their CRM results. By a factor of more than ten, "somewhat satisfied" executives outnumbered those who were "very satisfied." The causes for concern, even among the satisfied, were the implementation itself: resistance to change, integration with existing systems, and the cost. Other complaints included a failure to agree on the objectives of the system, and a concern about commitment by senior management. Forrester Research noted that the best CRM implementations focused more on the adoption than on the technology itself. The researcher calls this approach "synchronized deployment," with three distinct disciplines: process engineering, digitization, and user engagement. Process engineering is the link between people and processes. Digitization integrates the service provider's business processes with the CRM technology. User engagement links the people in the organization with the CRM technology.

One implementation challenge facing longtime systems developer AT&T was the variety of large legacy systems collecting and maintaining needed data for integrated customer management. AT&T hoped to reduce the time it took to find needed customer information while a customer waited, a delay that once took up to 15 minutes. A million customer records stored across three legacy applications, combined with the need for service representatives to understand how to use each disparate interface, contributed to the delay. AT&T was able to integrate data access with a single view to the user, without the need to rewrite the underlying applications.

Legacy systems present an implementation challenge for any large integration initiative, creating a benefit for new entrants, which can deploy the newest and most cost-effective technologies to meet information requirements. MCI entered the long-distance market with relational databases against AT&T's often-rewritten first-generation code. With the flexibility of the newer technology, MCI could support innovative billing practices and primeval but worthwhile customer relationship management support, making it impossible for AT&T to retaliate.

The debate over the value of CRM automation has not kept the industry leaders from participating actively in its implementation. Fujitsu Consulting surveyed 45 service providers and researched eight case studies from leading service providers such as Sprint, WorldCom, BellSouth, Verizon, and AT&T. The consultant found that 38% of survey respondents had completed a CRM initiative, primarily seeking lower costs. Most service providers maintained as their goal a special interest in channel optimization, to reduce call-center contacts as the percentage of total customer interactions from 45% to 31% in a 2-year period. Coupled with the introduction of best practices, most service providers claimed or expected to achieve acceptable cost reductions from their efforts.

Assuming that the service provider has both the funding and the stomach for the large-scale CRM automation commitment, the reward for success is the often-extolled 360° view of the customer. At first, this image evokes an existential nightmare from the customer's point of view beneath the high-intensity scrutiny of someone whose goal is to take more of the customer's money. Actually, well-constructed CRM systems that view the customer from every angle can benefit the service provider and enhance the customer's experience. This around-the-customer view is depicted in Figure 12.2.

Virtually every aspect of the customer relationship can find support in a well-integrated customer relationship management system. Sales force automation is among the most popular components of CRM systems, and the one that

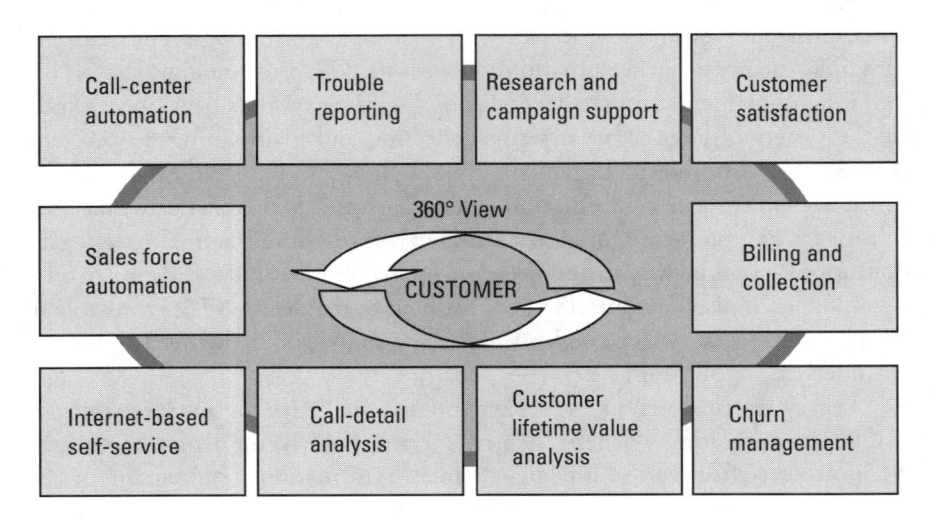

Figure 12.2 CRM process support.

even the most reluctant implementers praise. Call-center automation enables the service representative to retrieve all of the customer records in moments, communicate interdepartmentally to resolve problems while the customer is still on the line, and activate scripts to address routine customer requests while maintaining a façade of spontaneity with the customer. Customers can initiate trouble reporting whenever they choose, track the progress of problem resolution, and feed the customer profile for later data mining and analysis. Existing customers and their histories can provide research for potential segmentation and future customer acquisition campaigns to gain new customers. Interactions initiated by customers can launch customer satisfaction surveys, either to test overall satisfaction or only the encounter just completed. Satisfaction surveys can entice customers with prizes, coupons, or bonuses in exchange for their participation. On-line payments offer convenience to the customer and collection profiles to the service provider. CRM support for churn management can identify the moments when the customer is most vulnerable to churn, seek out the best channel to reach the customer and the most effective enticements not to leave, and create records to improve retention with the individual customer and a similar behavioral profile of future potential churners. Customer lifetime value analysis can review the customer base on the average for a snapshot of profitability. Furthermore, the CRM system can analyze a single customer's transaction history to determine the customer's anticipated ongoing profitability, and to decide whether to invest, harvest the account, or actively encourage the customer to leave or change account types. Telecommunications service providers are endowed with the capacity to analyze the individual customer relationship at the most granular level of detail, the calling pattern. Such an analysis can benefit both parties to the relationship and improve customer lifetime value. Last, many customers consider Web-based self-service an advantage, and because of its potential to reduce cost, service providers are happy to provide these capabilities to their customers.

12.2 Using Billing Systems to Enhance Customer Relationships

Billing systems are invaluable as a window to the customer for two reasons. First, billing enables the service provider to interact with a subscriber each billing period, and as long as the customer still believes that the exchange of money and service is a positive one—and that the billing process itself is not intrusive or inconvenient—the relationship will benefit. Second, the customer's call-detail record is a trove of valuable and proprietary marketing data that the service provider can analyze to strengthen the bond between the service provider and the customer.

J.D. Power and Associates rank billing among the six key performance issues that affect customer satisfaction. The convenience of a single bill is among the major enticements attracting customers to service bundles. According to Walker Information, only 46% of telecommunications customers are positive about the billing process. ISPs scored the highest of all the service provider sectors, though the rankings were not far apart, and no service providers ranked higher than 49%. The fact that customers notice their service provider's billing performance proves that it can be a competitive differentiator, but not necessarily in a good way. In fact, consultant Peppers & Rogers Group claims that 70% of calls to the service provider's contact center are billing related. Some 30% to 50% of business-to-business invoices have exceptions or disputes requiring human intervention, which delays payment [2]. Service providers with processes that generate such a high error rate and customer inconvenience, let alone dissatisfaction, would benefit from an evaluation of pricing structures, billing processes, and any other business function and should assign this review a very high priority. For those transactions that cannot be avoided, service providers can at least influence customer satisfaction by improving the call-center experience, or by eliminating the need for most billing-related calls entirely by ensuring that the bill is accurate and easy to understand.

Market researchers Lansdowne Market Research and Nielsen Media Research have independently found that customers who spend 15 seconds glancing at a piece of direct mail will peruse a bill for 43 seconds [3]. This offers the service provider an opportunity to reach out to the customer within the regulatory and practical limitations of the paper bill.

For those customers managing their accounts on-line, customers can often choose between automatic payment by debit or credit or manual review and payment. The automatic payment option is convenient for the customer and reduces or removes the problem of uncollectible payments for the service provider. For those less trusting of automatic payment, the monthly bill review interaction can present promotions tailored to the customer's account profile and usage patterns.

Service providers can enhance their on-line billing interfaces to provide competitive differentiation. IP-based local service provider Vonage requires billing through automatic credit card payment, available for on-line review. Innovative features include the customer's ability to change telephone numbers or service plans, view the entire service history on the Internet, and sort the reporting fields to meet individual desires. Telecom New Zealand touts its on-line billing service as environmentally friendly and enables customers to sort their charges, download charges to a spreadsheet for further analysis, view 6 months of service history, and perform a reverse number lookup when the customer fails to recognize charges for numbers on the call list. SBC sends the customer an e-mail when the bill is ready. Telus Quebec customers can click on an individual charge, such as an unrecognized call, and create an e-mail request to a customer service representative for an explanation.

The fact that some customers choose to conduct their billing activities on-line and others choose traditional channels might open some segmentation opportunities. Media monitoring service Mintel found that less than a third of adults whose service provider offered on-line billing participated in the service. The researcher further discovered that less than half of Americans who use the Internet were aware than on-line billing was available to them. As on-line billing benefits the service provider in reducing costs and providing a more intimate interaction with the customer, this gap between Web-enabled customers and Internet billing demonstrates a failure on the part of the service provider to educate its customers. Some customers know about on-line billing and choose not to use it. More than half of the respondents would be more receptive to on-line billing for a discount, an argument that should appeal to service providers, because they could pass along a percentage of the realized savings. Of those who choose not to go on-line, 54% of consumers say that they want the paper records to file, demonstrating another failure of the service provider to educate customers, most of whom undoubtedly have computers equipped with printers. The proactive service provider could also arrange to send a paper receipt after the transaction is complete to customers without access to printers or faxes, gaining most of the savings and retaining the customer intimacy and convenience of on-line billing. The problem of a paper receipt is of higher concern to those aged 65 and older; two-thirds of this group wanted to have the paper record, but this segment is least likely to have Internet access anyway. Innovative strategies to address the specific needs of this segment while maintaining low administrative costs could serve as a competitive differentiator.

In the aggregate, customer call-detail records can evaluate the impact on calling behavior of pricing changes, marketing initiatives, and network performance. For market segmentation, call-detail records can isolate segments that would be receptive to campaigns targeted to the destinations they call, the times they call, or the type of calls they make. Individually, call-detail analysis provides insight into the customer's preferences, whether they are receiving value, such as by using their plan's allotment of minutes, and whether they create or destroy value on a going-forward basis. Erratic call-detail behavior can also provide early detection of fraud against a customer's account.

Call-detail analysis provides a level of granularity and subtlety that surveys and other satisfaction devices cannot deliver. It measures what a customer does as opposed to what the customer purports to do. This analysis does not replace satisfaction research, but it augments other research efforts by providing an early indicator of customer future behavior. Service providers who are serious about reducing churn might offer customers lower margin packages that meet their actual usage patterns rather than risk defection to a competitor with an equivalent offer and less to lose. Call-detail analysis has its critics, including some of the customers under review. As call-detail data becomes more available to interdepartmental users and channel partners, and as its use expands, privacy will become more important to customers.

12.3 Customer Self-Service

Most organizations that target consumers, and many that target small business, view customer self-service as an important element of the customer care experience. More than three-quarters of call-center executives surveyed by Forrester Research consider self-service a "very important" or "critical" priority. Self-service is almost always Internet-based and can include a list of frequently asked questions, automated e-mail responders and scripts, natural language searches, a customer knowledge base, customer portals, or message boards.

An updated page of frequently asked questions can eliminate most of the routine requests directed to call-center personnel. Automated e-mail responders can add or delete customers from contact lists, send a statement of order status to the customer's mailbox, or perform other routine lookup functions. Scripts can find retail shops, inform the customer when new technologies will be available in their location, or calculate the cost of a customer-designed service bundle. Natural language searches using well-known engines can help customers navigate directly to the page containing the information the customer seeks. Knowledge bases offer technical support to the knowledgeable customer. Customers can access the same databases that the call center uses to answer questions from customers. Portals enable customers to manage their accounts in a secure and private environment, purchase services, and check on the status of orders, repairs, or other transactions in progress. Lastly, message boards create user groups for customers to share their experiences with other customers or present technical problems to support personnel monitoring the boards. Message boards can highlight the problems certain customers have with the service provider, though. Negative comments from vocal unhappy customers, coupled with the potential loss of control over the site content, have probably frightened away some service providers from using message boards. Moreover, there is undoubtedly little to share about creative uses of dialtone. As services become more complex and broadband content more common, message boards might provide a new channel for service providers to tap into creative customer-designed applications that the service provider can then develop into mainstream offerings.

Customer self-service is one of the few automation efforts that service providers and customers most often greet with glee. Yankee Group research during 2003 identified self-service benefits to the customer to include accurate and timely information, increased accessibility, improved business solutions, integrated solutions, automated troubleshooting, increased and faster access, targeted problem resolution, and targeted product and service information.

Though that list of benefits might well have been on the service provider side, The Yankee Group went on to list benefits to the service provider. These include increased customer satisfaction and retention, increased intelligence about customers, more selling opportunities and revenue, lower offline interactions and administrative costs, reduced call-center costs, improved process integration and efficiency, competitive differentiation, and consistency in the level of service.

Consultant Bain & Company has suggested that moving standard customer requests to the Internet can reduce the average cost per call by 97%. Bain estimates that about half of the calls to call centers are standard queries that can be answered on-line [4], saving the typical European incumbent service provider about 5% to 10% of operating costs. The primary benefit of migrating the customer to the Internet, though, is customer intimacy, not cost reduction. Personalizing the customer's relationship with the service provider creates an emotional link that is more resistant to churn than most other bases of competition. Moreover, the one-to-one experience of customer self service enables the service provider to customize the portfolio, the pricing, or the experience depending on that particular customer's value.

Fujitsu Consulting, in conjunction with software developer Netonomy, learned that the wireless customers they surveyed were most likely to use self-service customer care during evening hours and weekends. Besides the mediating impact of moving the traffic to underutilized network hours, self-service that replaces live customer service at those times produces greater savings. Call-center labor costs are highest during off-peak hours. The same study found that customers preferred self-service to contacting a call center or visiting a store for many routine functions such as reviewing bills, adding services, or changing service plans. Nearly 40% of those interested in self-service stated that they would visit the site at least once a week, providing exposure and marketing opportunities. As contract subscribers were especially interested in self-service, offering the self-service channel is a potential weapon against churn among particularly desirable customers.

Moreover, self-service is not only the cheapest delivery channel but the best one for certain customer transactions, such as detailed technical specifications or support, complex customer-led product comparisons, and transactions that interact with the customer's computer operating system, such as software or content downloads. Self-service is also likely to provide the most current information to the customer, because information can be uploaded to the service provider's domain as soon as it is available, as opposed to significant technical or marketing changes that require lead time to schedule and conduct the training of sales, call center or retail store personnel.

AT&T Wireless was among the first providers of on-line self-service, beginning in 1998. Besides lowering costs, the service provider found that some customers demanded self-service solutions, and that self-service customers tended to churn less than others. Less churn and lower costs meant that customer lifetime value for the self-service segment was improved as well. AT&T offered new rate plans to on-line customers tailored to their own usage, a feat that would be impossible using mass media and retail channels.

Self-service on the Internet is cost-effective, but it is often accompanied by transactions in costlier channels. Customer-initiated transactions can lead to deferred transactions that involve a labor cost, or an assisted transaction. Deferred transactions occur when the customer initiates the transaction through an e-mail or a customer contact form on the site and the service provider's response returns later in the form of a return e-mail, a callback, or some other contact initiated by the service provider. Assisted transactions involve a call-center representative or technical assistance, in the form of text chat, voice, or video chat in real time. Furthermore, as service providers succeed in migrating the most routine transactions or easiest technical support to the Internet, the average request of a call center will increase in complexity. Call-center costs per call could increase due to increase length of call, increased technical training, or the increased likelihood of supervisor involvement.

Self-service presents other challenges to the service provider. Some customers do not have Internet access or simply want to communicate with a call center. Some customers try self-service and fail, so they call for assistance in navigating the Web-based customer care, thereby occupying two service channels at once, neither one of which is working on the problem that prompted the transaction in the first place. The service provider employee on the receiving end of a customer e-mail might take so much time deciphering the request that the call center would have been more efficient. Nonetheless, these problems are largely temporary.

Other problems can arise when the service provider that the customer sees is inconsistent across channels. Customers will be dissatisfied if the transactions generated during the on-line session do not permeate the enterprise information systems and the business processes necessary to resolve or complete them. The customer who endures an unpleasant self-service session or a session that promises and then fails to solve the service problem will be very reluctant to trust self-service again.

Inquires that remain unresolved during the customer session can become costly and problematic. Gomez Advisors audited dozens of on-line retailers and found that consumers waited for a median duration of 12 hours for replies to questions, and in the end, only 40% were answered accurately. Telephone-based staff responses were accurate a better, but unimpressive, 63% of the time. The discrepancy between the two channels and the lack of reliable answers even from staff create a strong case to establish a Web knowledge base with validated answers to likely customer questions.

Aspect Communications conducted a survey of the 2003 holiday shopping season and found that customer service experiences had improved measurably since the previous year. Customers reported that e-mail accounted for 70% of their inquiries, and that they received replies to 20% of their inquiries within an hour. More than half were answered within a 12-hour business day (or nearly half were not, depending on whether the retailer or the customer is judging).

Because self-service is voluntary, service providers need to find ways to coax customers to go on-line. The customer segment that prefers on-line relationships is smaller than the candidates targeted for low-cost service channels. Internet-only deals, or offers with short expiration dates can entice customers to visit the Web site. But for customer service, the service provider needs the customer to choose Internet-based service whenever the need for it comes up. Aggressive service providers can put their Internet address on all customer communications. Service providers can send status notification e-mails only to customers who initiated the transaction on-line, and offer bonuses for making use of on-line support, sharing some of the cost savings with the customer. Additionally, the service provider can promote the Web site while customers are on hold for telephone-based customer care, though that gentle reminder might only exacerbate customer dissatisfaction with telephone-based care.

On-line communities are on the rise, and they help the service provider strengthen the brand with customers. Though most telecommunications service providers offer few complicated, customized or even technologically advanced services, on-line communities will grow in importance as these services come into existence. Customers can form strong relationships with brand managers, technical support representatives, and other customers. Communities give customers a reason to continue to visit the service provider's Web site, where the hosting service provider can expose them to marketing messages, within reason.

12.4 One-to-One Marketing

The ultimate application of customer segmentation in data mining and database marketing is to divide a universe of customers into compartments so small that each contains only one customer. This notion of particle marketing, or one-to-one marketing, was imagined decades ago but put into place only after technology had the scale to handle its volume. In theory, each customer undergoes a completely personalized experience in dealing with the service provider. Within the boundaries of customer care, one-to-one marketing implies that low lifetime value customers might be excluded from certain service channels, that high-value customers might enjoy very short wait times when contacting call centers, or that all customers will receive highly customized interfaces when interacting with the service provider. High-value (high-volume) business customers, with assigned sales representatives and a support staff in the back office, have traditionally been the only beneficiaries of one-to-one marketing. Technology has enabled service providers to offer similar treatment to consumers.

Telecommunications service providers are in exactly the right business to provide one-to-one marketing. First, for the most part, customers contact the service provider using the very account about which they would like assistance. Voice customers call operators by using their own lines which terminate in equipment that can gather caller ID information and pass it to a customer relationship management system's database in time for the service representative to take the call. Second, the industry collects so much customer behavior information in the course of business that only analysis is needed to produce highly customized profiles.

Nonetheless, the deluge of calling data represents only an unorganized pile of what the customer has bought. The more valuable knowledge is why the customer bought in the past, and what else the customer might buy in the future. The service provider needs to identify for each customer the processes with which they are most comfortable, and present them the interface that they value the most. For example, some customers prefer the lowest cost channels, such as Internet account management or automatic payment by credit card. High-value customers can receive preferential treatment, even through the Internet portal, through the use of content management technologies.

For customers who value human interaction, the service provider can route incoming service calls by an algorithm. High-margin customers can move to the top of the queue, while low-margin customers might warrant further analysis. Those customers whose billing cycle just ended will merit a different anticipated call length and skill set than those who just installed a new technology. Intelligent customer relationship management systems can apply the right skills to the task. Customers expect and should receive a continuous conversation with the service provider, so that they do not need to explain the history of their service relationship each time they connect to a new service representative or interact through a different support channel.

Service representatives can minimize the call duration for customers with a history of frivolous complaints and low lifetime value measures. But service providers need to live with their one-to-one marketing decisions. A 2002 study conducted jointly by Kelly Services and Purdue University's Center for Customer-Driven Quality found that 63% of customers will stop doing business with a company after only one poor customer service experience.

One-to-one marketing, at its logical limit, raises privacy concerns among some customers. According to researcher IDC, 37% of consumers have abandoned Internet transactions two to three times because of privacy concerns. Service providers striving for intimacy with their customers need to demonstrate that they are responsible with the information they capture and maintain.

According to the researcher The Customer Respect Group, 17% of firms in the telecommunications sector did not post privacy policies on their site explaining how they use personal data. This failure is especially inexcusable among telecommunications service providers, whom the public expects to assert leadership in matters of on-line responsibility. Service providers can be proactive about the ways they use customer data beyond the standard privacy agreement. By customizing the interface, they can generate a comfort level with marketing initiatives. An upfront indication of the use of known data about the customer might state, "Since you already have X feature, you are eligible for a discount on Y feature." With restraint and innovation, service providers can transform customer privacy concerns into a sense of security.

References

[1] AMR Research, "How You Define CRM Success Depends on Who You Are," January 2003.

[2] Robinson, Will, "Online Self-Service: Improving the Bottom Line and Pleasing Customers," *Telephony*, Vol. 243, No. 4, 2003.

[3] Hankins, Michelle L., "Using Telecom Bills as a Marketing Channel," *Billing World*, Vol. 10, No. 2, 2004.

[4] Matthies, Gregor, "European Telecoms Should Dial Up Web Savings," *Global Telecoms Business*, Issue 58, April 2001, pp. 40–41.

13

Branding

13.1 The Importance of the Brand

The customer-centered service provider recognizes that the image of the brand rests with the customer. Service providers can invest persistently in a brand to shape its impressions, but in the end, brands exist only inside the minds of customers and other stakeholders.

Historically, telecommunications service providers have been inconsistent in attaining success in branding or positioning brands. Monopolies enjoyed strong overall brand awareness if not product brand strength, but over time the brands retained attributes that service providers in competitive markets do not desire: high price without value, lack of a customer focus, lack of progressiveness. New entrants such as the United Kingdom's Orange were able to start out without the preconceived notions by customers. Unburdened by history, Orange could invest in a comprehensive branding and positioning program and capture a share of the dramatic growth in the wireless market at a faster rate than its competitors did. In the same geographical market, the incumbent service provider British Telecom struggled to reposition its multiplicity of brand names and conducted a structural separation and had less success in generating brand excitement. Similarly, in the United States, AT&T Wireless was unable to create a strong image with its m-Life campaign, though it was similar in focus to the more successful lifestyle marketing of Orange in the United Kingdom and Cingular and T-Mobile in the United States. Especially in wireless services, branding is imperative for attracting customers, because without strong brands, customers will select service providers on the basis of price alone. Branding efforts have been laudable in wireless markets around the world, yet Yankee Group found that 85% of European mobile users surveyed indicated that price was a key reason to switch service providers. Brand was the second most important factor.

The marketing function controls most of the venues through which brands are developed and nurtured, as depicted in Figure 13.1.

Advertising is the best-known route to developing and nurturing the brand. Verizon Communications was the seventh highest advertiser of any kind in 2002, spending more than a billion dollars U.S., and in the first three quarters of 2003, spending almost $900 million, according to TNS Media Intelligence/CMR. Advertising expenditures rise and fall with economic times and market shares. In 2002, the top two U.S. spenders were AT&T and Verizon, according to the Advertising Age Megabrand Report. Spending does not guarantee success; in a Business Week/Interbrand survey in 2002, AT&T, seventeenth on the list of all brands, had lost 30% of its brand value in a single year.

Service quality is part of the brand image. Not only is service quality a fundamental component of the customer's experience, poor coverage and other indicators of quality are a significant cause of churn. Though the marketing function does not control the network itself, marketers can provide important feedback to network planners to improve the customer's experience and perception of the brand. Marketers can also use advertising and other branding venues to swing customer opinion by connecting the brand with a positive image of quality, as Sprint did with its long-running "pin-drop" campaign and Verizon reiterated with its "Can you hear me now?" slogan.

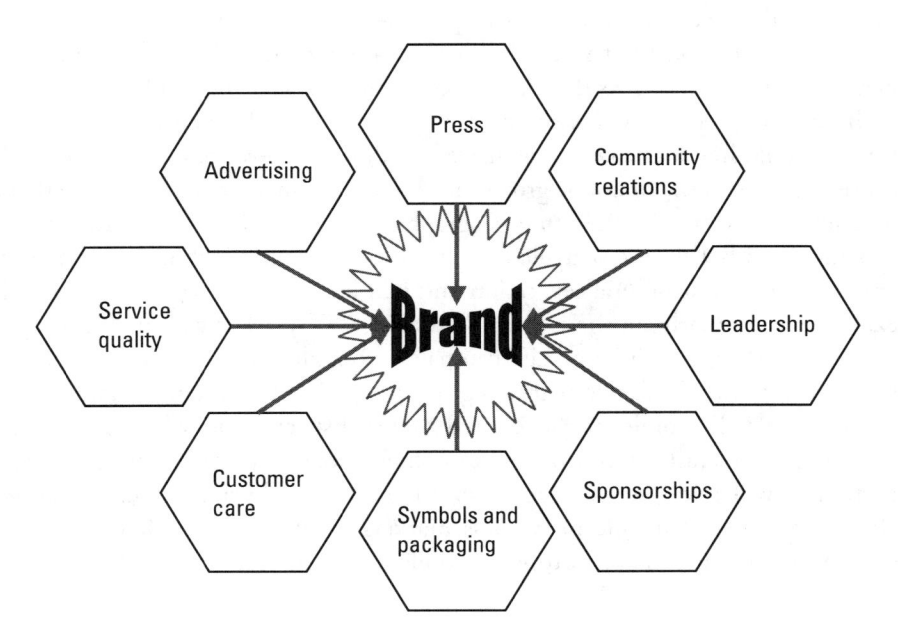

Figure 13.1 Marketing-controlled brand management.

Customer care can affect the strength of the brand through all channels where customers interact with employees or anyone representing the service provider, including technicians, agents, and all channel partners. Internet self-service, where innovations on the portal differentiate one service provider from its competitors, have the potential to increase the future value of the brand. An independent survey sponsored by software developer Genesys found that 80% of its respondents said that customer service representatives have a significant or a major influence on their opinion of a company.

Service providers use symbols to convey the brand image to consumers. Canada's Telus mobile service provider has used lizards, frogs, ducks, ladybugs, monkeys, and a pair of Vietnamese potbellied pigs to portray itself and brand its services with friendliness, which the service provider believes is a cornerstone of its brand. Omnipoint used a parrot whose communication was so clearly transmitted that it did not appear to occur to the company spokeswoman on the receiving end of the call that she would be dating a bird that evening. Cingular uses an animated child's jack, and several other wireless service providers use animated or fanciful symbols in an effort to convey the friendliness and spontaneity of wireless service. Not only does a symbol or animal limit the expense for on-camera talent, the brand is safe because spokescreatures never get arrested. Orange makes a strong statement with its orange-colored palette. Many wireless service providers favor the colors yellow, orange, and red in their collateral materials, conveying friendliness, energy, and boldness.

Service providers align themselves with not-for-profit organizations, sports, or other affinity groups to capture the attention of desired market segments. Sponsorships also enable the service provider to leverage the attributes of the event or group sponsored. Sponsoring a sporting event creates a particular aura for the audience of that event; sponsoring the arts reaches a different segment and can evoke a different image. Nextel incorporated its NASCAR auto racing sponsorship into a series of proprietary handsets. Sprint has set up hospitality tents at golf tournaments, giving the service provider targeted access to high-end consumers and business customers. One Cingular commercial used its sponsorships—first as a parody, then as a targeting tool—to show its commitment to sponsoring competitive events, and secondarily to demonstrate the importance of selecting the right partnership.

Company leadership can play a major role in putting a face on the brand. Virgin Mobile's chief executive Richard Branson is recognizable for creating the Virgin conglomerate and for his highly visible worldwide adventures. Virgin's no-contract policy and straightforward service and payment plans fit its other corporate ventures. This path to brand enrichment appears underutilized, according to a 2003 study by Opinion Research. Only about one-quarter of surveyed individual investors said that corporate chief executive officers cared enough about their company's reputation. Noninvestors showed similar results.

Nonetheless, close association between a CEO and the brand risks irreparable brand damage should the CEO resign or lose credibility for business-related or unrelated transgressions.

Service providers can nurture their brands in the local communities they serve through community relations programs. Verizon promotes literacy through its Verizon Reads program. Most telecommunications industry leaders promote causes, practicing and then communicating their social responsibility commitments to influence brand image. Like all brand influencers, community events can have a positive or a negative effect on the brand image, directly or indirectly. Labor or other local problems in one community can cause negative publicity and influence the brand image in other communities.

The press can assist in conveying brand impressions, for better or for worse. Press conferences and press releases at the time of a service launch for a new brand can leverage other marketing expenditures by generating publicity at no cost. Though these efforts will undoubtedly create news, service providers have little control over how the news is presented to the public.

13.2 Brand Equity

Though the brand itself is invisible, it is somewhat measurable in the form of brand equity. In short, brand equity is equal to the value of the brand above the unbranded value of the underlying assets. Therefore, brand equity is a measure of the relationship between a service provider and its customers, and an indicator of how likely the customer is to continue to buy. If a service provider can sell its wireless minutes for a certain price, and an unbranded reseller sells minutes at a lower price on the same network and using the same support structure—without alerting the customer to the source of the service—the premium that the branded service provider receives is evidence of brand equity. Service providers invest millions, even billions, every year into efforts to create and maintain strong brands. High brand equity reduces churn and reduces the cost of marketing new services to existing customers. Higher margins on services with identical costs means that brands with positive equity are much more profitable than their competitors. Strong brand equity results in customer recommendations to new customers, adding market share at a minimal acquisition cost.

The constant rebranding that telecommunications services have undertaken implies that few brands in the telecommunications market have had sufficient equity for their management teams to retain them against mergers with other brands. A court battle between AT&T and the divested regional companies two decades ago left the local service providers with the Bell name and logo, but today's American landscape holds few reminders of the name's former power. AT&T's corporate headquarters kept the name, but the broadband bearer of the AT&T name disappeared into that of its acquirer in the Comcast cable merger, even though the AT&T customer base was larger than that of its buyer. The AT&T Wireless name is also destined to pass into history. The wireless division's initial spinoff and subsequent decline in market share provide some evidence that tight integration with the parent's brand was not the ideal marketing approach. MCI's strong name nearly disappeared into its acquirer WorldCom, only to be resuscitated after a scandal and the subsequent hunt for an untarnished identity.

Marketing consulting firms disagree about the components of a brand equity measurement model, but two quick tests can enable the service provider to ascertain whether brand equity exists. The first test is the difference between the assets on the balance sheet and the market capitalization of the firm. Telecommunications services are asset-intensive, and will probably never achieve the 90% equity and 10% assets ratio enjoyed by branding leaders such as Coca-Cola. On the other hand, at any moment, service providers can calculate that ratio for themselves and about their publicly held competitors using readily available data. At the very least, the service provider can reveal relative market positions, if not absolute value. While this test does not evaluate customer beliefs, the stock price reflects investor opinions of customer intentions and can act as an easy-to-compute surrogate.

The second test is slightly less objective, but demonstrates a more direct market link to customers. A comparison of the pricing for virtually identical service packages will provide some indication of the relative brand strength of each package. This comparison is more finely tuned than an analysis of market capitalization, because it shows the brand strength of a single customer offer. For example, unbranded bundled packages of unlimited local and long-distance calling in the United States typically sell to consumers for prices at least 20% below those of the market leaders.

The most successful brands deliver to the customer's present expectation level or better. The visual expressions of identity are consistent across media, across channels, and across services. Brands express core values and help to set future customer expectations. The customer who is bombarded with advertisements extolling network quality will be more disappointed with dropped calls than the customer whose expectations for low price are met.

It is one thing to track brand equity and yet another to manage it. Tracking involves testing the attributes of one's own brand against the attributes of competitors, a result that the service provider can then plot on a perceptual map. Of those attributes, some, such as customer service at any hour, are baseline requirements simply to remain in the marketplace. Other attributes will be neutral, neither attracting nor repelling customers. The attributes that differentiate the service provider but do not appear to matter to customers will not contribute to brand equity either. Service providers should revisit neutral attributes that incur investment or expense but add nothing to competitive differentiation. Negative attributes require immediate attention, though negative attributes that do not vary among competitors are a lower priority. On the other hand, the service provider cannot simply ignore negative attributes that do not vary between competitors; the service provider that eliminates the undesirable attribute will steal market share quickly. Eliminating roaming and long-distance charges catapulted AT&T Wireless into a leading service provider within months of its market entry. Eventually, flat-rate local and long-distance minutes regardless of originating locality became a baseline requirement for all U.S. wireless service providers.

Brand equity management involves finding the emotional connection between the customer and the brand. For a wireless consumer, it might revolve around the handset. Suppose the customer only uses the wireless phone away from home and office, in the presence of friends. A very modern handset or one with special features might make the customer feel wealthy or gain a sense of status. The service provider can build on this emotion by adding features that foster this emotional state.

13.3 Product Positioning

Classical brand management usually applies to products rather than services and to corporations whose portfolios contain dozens or hundreds of distinct brands. Telecommunications service providers have only recently been successful at positioning individual components of their service portfolios, emphasizing their overall image instead.

Service providers stake out branding positions in a context of other service providers. The most common positioning strategies—high-quality networks, low prices, or excellent customer service—make the most sense when compared implicitly to the inadequacies of competitors. These popular strategies correspond well to the value disciplines of product leadership, operational excellence, and customer intimacy described in Chapter 3. Verizon Wireless and Sprint have pursued brand leadership in the area of service quality. T-Mobile in the United States has targeted the low-price seeker. Nextel has sought a business-customer base, a segment that is typically most interested in reliable service and responsive support. Nextel offers walk-in or mail-in repair service for fast turnaround, a corporate service program, and insurance.

Service providers that own the leadership position for one of these value disciplines need to invest in keeping that leadership. Service providers that hold second place or farther down the leadership ladder need to unseat the present leader or find somewhere else to lead. They can identify and target an underserved segment, create a niche, or reposition the competition.

Preemptive positioning occurs when a service provider uses an attribute as a differentiator even if the attribute is not unique to the service provider. Both Sprint and Verizon centered their branding campaigns around network service quality, though hard evidence from objective scientific appraisals that their networks were appreciably better than their competitors was at best not very well publicized to customers. Nonetheless, branding is in the mind of the customer, and very high customer ratings for network quality followed their campaigns, even when survey respondents had no first-hand experience using the networks about which they were asked.

Positioning through a product feature is common in most market segments. Typically, the marketer selects a feature that differentiates the service from its competitors, when customers might not realize its utility. For example, T-Mobile's introduction of the camera phone in the United States showed customers how to solve everyday problems with the feature, a utilitarian approach, even though the appeal of the handset for most of its potential buyers might have simply been its stylishness, an emotional appeal. A similar positioning technique is to use the benefit a service provides as a branding strategy, as did 1-800-COLLECT in the United States.

Head-to-head positioning takes place when one service provider attacks a single competitor on similar product attributes in the interest of targeting the same market segment. MCI used this strategy, using price as a differentiator among otherwise equal competitors, as a cornerstone of its attacks against AT&T, even when there were thousands of low-priced resellers in the long-distance market, and even when Sprint's market share was closer to MCI's than MCI's was to AT&T. This strategy only worked because the industry was undergoing deregulation, prices were falling rapidly, and call volumes were growing. Most importantly, AT&T was reluctant to batter MCI with competitive retaliation in its ongoing antitrust scrutiny at the time. Astute small service providers will only use this technique in a normal competitive market when the nearest competitor has similar market share.

Exclusive club positioning uses distinctive membership to differentiate the brand. For example, promoting one's qualification in the Fortune 25 or as the company of the year in some distinguished category enables the service provider to differentiate itself from its competitors.

Superlative positioning claims to be the best, the largest, the oldest or some other superlative that appeals to prospects. Some countries prohibit this method of positioning. Furthermore, it would be inappropriate in many telecommunications markets, where incumbent service providers would rather minimize references to their former monopoly status and persistent retention of market share leadership.

Indeed, for a service provider in transition from monopoly to competition, the challenge of repositioning the brand in the customer's mind is crucial. This repositioning of the mind might require as much effort as repositioning the service provider's own business processes to function in the competitive environment, though both efforts are necessary. Consulting firm McKinsey suggests that many repositioning efforts fail because the service provider fails to distinguish between aspirational and achievable positioning. Service providers that promise, for example, full-service telecommunications create customer expectations for delivery. The service provider that is limited geographically, or that forces customers to maintain multiple service channels, or that retains vestiges of antiquated business processes will fail to reposition successfully. According to McKinsey, attempting to reposition a brand too far from the customer's frame of reference regarding the original brand position will confuse the customer and harm the brand. Consequently, the incumbent service provider being attacked as too expensive by its discounted rivals might abandon its goal of being the low-cost provider and strive to position itself as the high-quality brand. Service providers need to consider their own measurable quality as well before selecting a platform for branding. Social scientists have identified a syndrome that they call "the Lake Wobegon effect" after radio entertainer Garrison Keillor's fictitious hometown. Organizations, including service providers, tend to self-report an inflated perception of their positioning in the marketplace, or as Keillor describes the Minnesota town: "where the women are strong, the men are good looking, and all of the children are above average."

Wireless service providers have also contended with rebranding as a result of industry consolidation. SBC and BellSouth branded Cingular from its inception as a service independent from its owners. Verizon eliminated the powerful but somewhat limited names of GTE and Bell Atlantic in favor of its Internet-friendly, evocative new word. Deutsche Telekom's T-Mobile gave up whatever brand equity it had in the names of predecessor companies around the world, including One To One in the United Kingdom, VoiceStream and Powertel in the United States, Czech Republic wireless provider Radiomobil, and max.mobil in Austria.

13.4 Branding Commodities

A commodity is a product or service that customers view as identical to its competitors. Incumbent service providers once feared that the telecommunications services they offered would become commodities, in other words, that customers would view them as completely interchangeable in the marketplace. This is a rational fear; the best way to compete successfully in a commodity market is through low price, and that results in low margins. Furthermore, monopoly service providers historically provided value and high service, attributes not associated with low price. Commodity marketing would have forced incumbent service providers to change their business processes and cultures even more radically than they did.

Ironically, the sectors that telecommunications service providers feared customers might have viewed as commodities—basic voice and long distance—are not only interchangeable, they have nearly disappeared as distinct services in the consumer market. The leading brands of long-distance service, AT&T, MCI, and Sprint, virtually stopped investing in differentiating their brands from competitors and instead invest in differentiating their service bundles or their wireless offerings. Long distance itself migrated to wireless packages, local service bundles, and will likely erode further into VoIP packages and nonvoice alternatives such as e-mail, instant messaging, and SMS. Fixed-line local voice services shrank in response to wireless substitution and service bundles that did not differentiate between local and long distance. Qwest's offer of DSL-only service without companion local voice services was the first to market its recognition that local fixed-line voice service had become a commodity that could finally serve as a differentiator in its own absence, not its presence.

Bandwidth became a commodity due to its interchangeability and its overabundance. A commodity service offers two marketing approaches: compete through operational excellence and low price, or find a way to differentiate the service and create a brand by eliminating its commodity status.

Internet-based shopping and buying has the effect of creating commodities where customers might not have access to them in other channels. Intelligent shopping agents, information-rich Web retailer sites, and doggedly determined searches place in-depth product comparison within reach of the dedicated Web surfer. These comparisons will undoubtedly become more accessible in the future. The customer that knows exactly what service bundle to buy, for example, can already price nearly identical packages from a universe of service providers in adjacent browser windows. Thus the service provider does not even gain the advantages from retail store convenience and the likelihood of shopper fatigue in competing against similar service portfolios.

Consulting firm Booz Allen Hamilton recommends a four-phase approach to branding commodity products [1]. The first step is to segment the market to identify whatever customers might be receptive to segmentation. For those customers who demand the lowest price, the consultant recommends refraining from additional marketing investment or even giving those customers to competitors as a backhanded gift. The second step is to create differentiation along any of six dimensions of differentiation. The six dimensions are quality control (service consistency), reliability (consistency of supply), packaging (product convenience), taking responsibility (delivery consistency), matching (customization), and knowledge-based applications. In a service provider context, quality control cannot simply mean a strong network connection, because that is not a differentiator. It implies a proprietary feature, such as demonstrably less latency than its competitors, as in Nextel's push-to-talk capability. Similarly, reliability might mean wireless coverage if one service provider clearly excels in a geographic territory, but that differentiator is almost certainly temporary. A better differentiator might be a service agreement for a business customer or a commitment to repair a handset at no charge beyond the initial service contract for a wireless consumer. Product convenience might manifest itself in innovative recharge mechanisms for prepaid services. Taking responsibility involves the delivery of goods, and is problematic for service providers, whose customers use services without asking for them. Invariably prompt provisioning could differentiate one service provider from another; self-provisioning capability might create a brand. The service provider can accomplish matching with customized bundles, and knowledge-based applications can occur when service providers learn about vertical markets and offer solutions to business problems unique to the customer's business.

In business markets, many opportunities exist for transforming commodities to brands. Business customers evaluate the total cost of their own value propositions, so service providers that use telecommunications solutions creatively to eliminate business costs in other areas, such as travel, can gain a portion of the savings. Consultant McKinsey evaluated the buying preferences of a customer of industrial resin, a quintessential commodity, and found that price accounted for only 30% of the purchase decision [2]. McKinsey identified three types of customers: service oriented, making up 20% of the total, product oriented, 35% of the total, and price sensitive, 45% of the total, in a market that many consider to be commoditized. Considering that only a small segment of customers are responsible for most of the profitability, McKinsey suggested that one-to-one interviews with customers selected because of their high lifetime value could unearth those attributes that can command investment in differentiation.

A statistical technique called conjoint analysis is well suited to functional brands like telecommunications rather than fashion brands because it is based on a rational customer decision process. Conjoint analysis measures the perceived values of specific product features, and can identify the demand for the product as it relates to price. This analysis can divide the customer base into those who are concerned primarily about price and those who might pay a premium for features beyond low price. The technique is especially accurate because it forces customers to choose between sets of attributes and then uses multivariate statistical techniques to identify which attributes are most important to customers.

13.5 Protecting the Brand

A list of best-known telecommunications brands, compiled by Corporate Branding LLC in 2000, demonstrates the difficulties of making brands endure.

The only manufacturer—irrelevant to this analysis—to make the top 10 list was in first place, as shown in Table 13.1.

Of the 10 top telecommunications brands identified as familiar and favorable by key business decision makers in 2000, only a few even remained in existence less than 5 years later. Most often, these brands were lost to spinoffs, mergers and acquisitions, whose new corporate parent either believed that retaining the brand name was not worth the investment, or that simply wanted a consistent face across all its brands.

The discarding of a brand identity newly purchased at a premium is puzzling, or will be at least short-lived, demonstrated by a formula that is slightly oversimplified. The brand equity or value is the difference between the asset value and the market capitalization (or approximate acquisition price). Therefore, buying at the price of market capitalization is more costly than simply replicating the assets of the purchased company would be. A starkly rational asset purchase would simply duplicate the acquisition target's assets and compete using the brand of the acquiring company. In the short term, mergers in the telecommunications services industry have provided nationwide coverage to regional companies, transferred ownership of spectrum or jurisdictional licenses, offered market entry to sectors once prohibited, or created a critical mass of customers for service providers to compete with industry leaders. Mergers already completed were apparently not seeking brand equity, though they were paying as though they were, because if they were trying to acquire a brand, they might have kept the brand they bought. A continued dearth of strong brands would

Table 13.1
Brand Restructuring, 2000–2004

Company Ranked in 2000	Status of Brand, 2004
1. Motorola	Not applicable; equipment manufacturer.
2. AT&T	Some branding disappeared from cable broadband, wireless, and equipment; wireless and equipment spinoffs experienced larger market capitalization than former parent.
3. GTE	Became part of Verizon.
4. MCI	Rebranded as MCI WorldCom, returning to MCI brand.
5. Sprint	Intact in 2004.
6. Lucent Technologies	Former AT&T brand, equipment provider.
7. Bell Atlantic	Became part of Verizon.
8. BellSouth	Intact in 2004, but associates its wireless with a different brand.
9. SBC	Intact in 2004, but associates its wireless with a different brand.
10. Ameritech	Became part of SBC.

hamper further industry consolidation once these nonbrand reasons become less important. Without brand equity as a merger motivator, service providers would be wiser to invest in growth through construction of new assets branded in their own, carefully nurtured brand image.

Protecting the brand requires more than routinely filing for patents and trademarks and prosecuting infringements. Most brand erosion is traceable to its owner. Protection requires visual and operational consistency within all business processes, training of employees to achieve that consistency and protection, and vigilance to guard against unreasonable line extensions.

Line extension occurs when a service provider launches a new brand by leveraging the favorability of the reputation of an existing brand. AT&T Wireless recovered from a late market entry partly because of its reputation as a telecommunications leader, though much of its rapid growth was also due to an innovative pricing structure. The AT&T brand was not strong enough to sustain the service provider in an enduring leadership position once its portfolio was largely undifferentiated from its competitors. The AT&T brand remained strong in the residual business of long distance in 2003, according to J.D. Power and Associates, but it shared second and third place with Sprint among high-volume users. Relative newcomer Verizon, containing the former GTE long-distance network but most strongly associated with local and wireless services, unseated both of these venerable long-distance providers to win first place in the long-distance survey. Verizon scored highest among high-volume users in the six attributes of overall satisfaction, performance and reliability, cost of service, billing, image, offerings and promotions, and customer service. On the single attribute of image, the best indicator of brand, Verizon topped the list, followed by AT&T, Sprint, and MCI.

Brand leadership might tempt Verizon to extend its strong brand horizontally to offer services or equipment that fit with its present portfolio, but there are risks to unrestrained brand extension. Not only is there no guarantee that the new services would be successful, there is the risk that unsuccessful extensions could corrupt the original brand's value.

It is easy to measure that a brand is strong, but not as easy to chart the exact territory the brand occupies in the mind. Suppose a service provider's image is strong as a result of good network service. That service provider tops the list of service quality ratings year after year. Then the service provider decides to extend the brand horizontally to some new technology as a first mover. If the new technology does not perform as well as its established networks—never mind that competitors' versions of the same new technology perform no better—the brand extension might not only dilute the brand image; it might even tarnish it. Service providers need to decide whether the risk of losing a long-cultivated asset is worthwhile, compared to launching a separately branded service with a steeper awareness climb, but a shorter distance to fall.

There are some indications that service providers, even the incumbents, are not only willing, but also eager to shed their long-term brand images in favor of untested but favorable prospects. Naming a wireless joint venture Cingular instead of some mishmash of its owners' names or naming a new Internet service provider Wanadoo instead of France Télécom Online enabled the rebranded companies to compete evenly and successfully in the same arena with newcomers. The choice to abandon the incumbent names also abandons the staid image of former monopolists, a plus in a marketplace that targets youth and value seekers with sunny colors and images of fun. Furthermore, once positioned, a branded service will be more durable than an entire company image when the customer is seeking a service to buy. Service providers will take their lead from consumer products companies that simultaneously support dozens of brands and save their corporate image advertising for the financial markets.

References

[1] Hill, Sam I., Jack McGrath, and Sandeep Dayal, "How to Brand Sand," *strategy+business*, Issue 11, Second Quarter 1998, pp. 22–34.

[2] Forsyth, John E., et al., "Shedding the Commodity Mindset," *The McKinsey Quarterly*, No. 4, 2000.

14

Marketing-Based Innovation

14.1 Distributed Innovation

"Because its purpose is to create a customer, the business has two—and only two—functions: marketing and innovation. Marketing and innovation produce results. All the rest are costs." So stated management guru Peter Drucker in his 1954 classic, *The Practice of Management* [1]. To paraphrase the aforementioned baseball umpire of Chapter 3, innovation is nothing until somebody markets it.

Innovation is often defined as a new product or service, a new process or method, or a new idea. The fuzziness of its definition does not keep service providers or anyone else in most industries from taking ownership of the concept. Innovation is good, by nearly every measure. More than 60% of executives polled in October 2003 by consultant Boston Consulting Group would increase their spending on innovation in 2004. Yet approximately the same percentage said that they were unhappy with the returns on their investments in innovation to date.

Continuous innovation refers to technologies that are new but have little effect on existing consumption patterns. Changing modem speed from 9,600 baud to 19,200 baud probably did not attract more new customers to the Internet than did the simple inclusion of modems in personal computers, though it was twice as fast (or half as slow). *Dynamically continuous innovation* is so new that it changes customer behavior. Broadband had this effect on Internet users. VoIP promises to offer *discontinuous innovation*, newness so dramatic that it changes the customer experience.

Innovation once began at home, in the laboratory, but like so much else, innovation distributes itself through the network of the marketplace. The customer-centered service provider needs to take a slightly slanted look at innovation. Like a strong brand, the brand value of innovation takes shape only inside the customer's mind. Furthermore, technology sectors have observed an innovation migration from the laboratory outward to the marketplace, as shown in Figure 14.1.

Innovation in the telecommunications services industry began as a top-down process. In the United States, AT&T conducted virtually all of the research and development across the telecommunications industry, including end-to-end equipment and the services that the equipment supported. Most service development was hardware-based. Independent networks like GTE and foreign equipment manufacturers added some innovation, but AT&T and its subsidiaries drove most of the innovation for nearly a century because the monopoly controlled most of the marketing channels and operating venues. When innovations related to long-distance calling were launched, AT&T and local operating company representatives trained not only their own employees but also the employees of unaffiliated independent service providers, and AT&T's nationwide advertising familiarized all customers with the new features. AT&T controlled the type of innovation and its pace.

As technology became more customizable and user friendlier, the need for centralized service development was reduced. By then, AT&T and its development laboratories were in pieces after the company's 1984 divestiture and the emergence of competition forced each service provider to differentiate through innovation. Marketing organizations of service providers could respond to the needs of unique segments or create spontaneous campaigns for holidays or other temporary promotions. Some service development became software based. The authority required for creating a localized or individualized service could

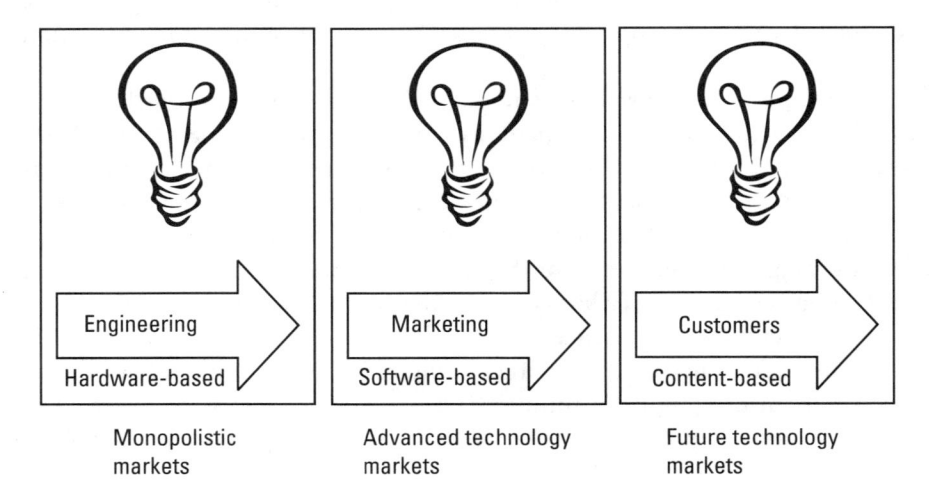

Engineering	Marketing	Customers
Hardware-based	Software-based	Content-based
Monopolistic markets	Advanced technology markets	Future technology markets

Figure 14.1 The migration of innovation centers.

become only a matter of company policy. Software-based pricing enabled service providers to create individualized bundles at a negligible development cost, because users could code pricing parameters into rules-based engines without the intervention of programmers. Marketers could decide at the last minute to run a promotional price on a specific long-distance route, or reprice minutes as a temporary feature, or add discounts to unique service bundles that were not part of the standard service portfolio.

The third wave of customized innovation will rest with the customer in the future. Services will become more content based. Hints of customer-developed services have already appeared in the computer games industry. A 2002 Booz Allen Hamilton survey of 94 game developers found that more than four-fifths of them sponsored on-line communities and more than one-third offered toolkits for customers to design their own product enhancements. This implied a future in which game developers present half-finished products to their user communities and let their customers finish their development.

On-line communities that have the potential to participate in service development take the form of newsgroups, bulletin boards, chat rooms, and forums. Predictably, the first on-line communities centered on technology, but today's on-line communities include business organizations, consumer products, and hobbies of every kind. Service providers will find opportunities and risks in sponsoring on-line communities, and when the third wave of innovation arrives, they will have no choice but to become the caretaker of customer-developed service enhancement.

14.2 Marketing Disruptive Technologies

Disruptive technologies that dramatically alter the landscape of an industry demonstrate the power of innovation. In the *Innovator's Dilemma* [2], Dr. Clayton M. Christensen demonstrates that the very traits that make industry leaders successful, indeed customer-centered, can cause their decline in the face of disruptive technologies. The author defines a disruptive technology as a technology that has the potential to supplant the existing technology. In contrast, sustaining technologies represent improvements to existing products. DSL is a sustaining technology, optimizing the copper loop to handle broadband transmission. VoIP is a disruptive technology, entering the marketplace uncertainly but likely to replace the existing infrastructure.

The reason that disruptive technologies are as important to marketers as to engineers is that marketers control their introduction, marketers evaluate their potential, and marketers are somewhat responsible for the damage that disruptive technologies do to established service providers. Disruptive technologies are most often developed in the laboratories of industry leaders, though they frequently come to marketplace prominence alongside new entrants. These technologies start out cheap but feature-poor, without a waiting market, and they tend to repel the current customers in contrast to the products or services they already use.

Disruptive technologies make good marketers into bad market strategists. The good marketer within the industry leader's organization listens to customers and responds to their needs. Customers tell the marketers that they do not need these new technologies. Good marketers respond to their competitors, who are listening to their own customers and eschewing the disruptive technologies. The good marketer invests in services with known markets and high margins and avoids service introductions that would cannibalize the existing portfolio. Customer-centered marketing is one of several reasons that good marketing is bad strategy when disruptive technologies come along. The paradox is that customers eventually decide which disruptive technologies succeed, yet customer-centered marketing rejects this type of innovation.

The relationship between technology and its sustaining and disruptive replacements are depicted on Figure 14.2.

Great marketers see the long-term value of disruptive technologies, and they learn to recognize them when they see them. Disruptive technologies address the needs of the present market, but not with distinction, and often less ably. They are frequently cheaper than present technologies. Technology improves along a curve shaped like the letter S, and much research has examined the technology S-curve. The curve begins in a shallow arc, and then rises steeply as the market absorbs the technology. Eventually the vertical growth abates, creating a flatter shape as the technology matures. Sustaining technologies overtake

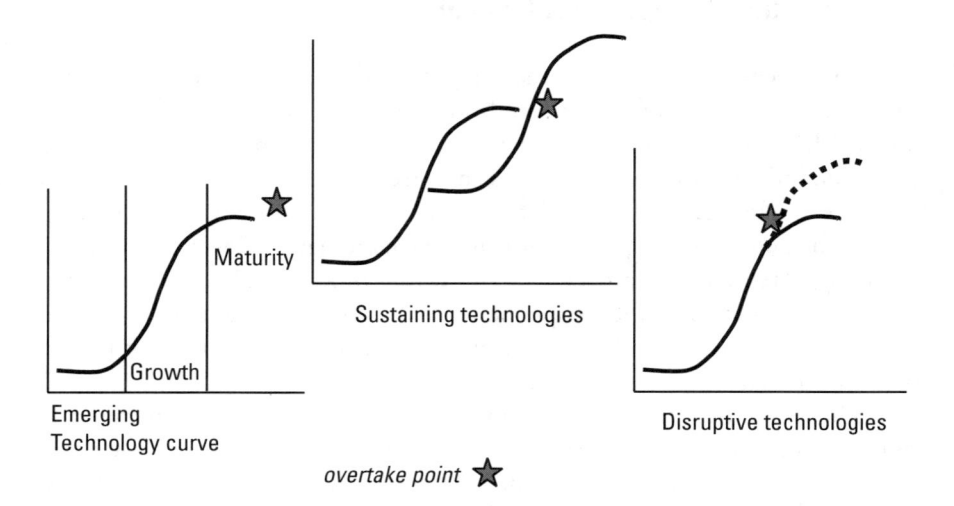

Figure 14.2 Disruptive technology marketing.

existing technologies during the maturity phase when the technology matures and the original technology's curve becomes shallower.

But the slope of the curve of the disruptive technology is steeper than that of the present or sustaining technology, indicating that the disruptive technology will overtake its predecessor's performance. Moreover, disruptive technologies are likely to overtake during the original technology's growth phase, while new customers are still joining the fold, and before the original technology has captured all of its anticipated profits. Service providers instinctively protect the proven technologies with additional investment, supported by customer feedback and competitor activities, but that approach might not be in the long-term interest of either the customer or the service provider.

The challenge is to how to identify which new technologies are truly disruptive, and to position them in the marketplace so that they are ready for customers when they realize they want them. Christensen identifies three litmus tests to determine whether a technology will prove to be disruptive.

The first test asks: does a growth opportunity exist? Disruptive technologies occur outside of the present served market or they offer a new business model to customers at the low end of the mainstream market. The second question asks: can the technology draw mainstream customers away from the present offering in spite of its limitations? Sometimes the mainstream customers are attracted to a technology even though its performance does not yet equal its incumbent counterpart. It took several decades before a significant number of customers substituted wireless service for their landlines, and the market grew by double digits before the substitution took place. It is not surprising that substantial wireless substitution began to occur at the point that the service itself (broadband and Internet capability) and its regulatory surroundings (local number portability) approached the standard of its fixed-line counterpart. The third question asks: can the incumbent respond? Though incumbents are often the developers of disruptive technologies, they have the most to lose from their dissemination. Margins on sustaining technologies are substantially higher, and in the telecommunications sector, disruptive technologies generally require capital investment and no small risk.

One signal that a technology is disruptive is that technology experts disagree as to its potential. Clearly that sign is not conclusive, though, as that group would include AT&T's disastrous Picturephone video service. Another attribute is that the new technology serves a market that was not served by the existing technology. A disruptive technology might have a performance improvement slope that is steeper than existing technologies, even if it presently underperforms existing technologies. But it is more important that the slope exceed the needs of the market rather than the existing technologies.

Many industry leaders have lost market share to disruptive technologies, even when they were developing the new technologies on their own. These failures were self-imposed, and organizationally endorsed. The typical innovation captures the low end of the market, or a tiny new market by targeting those who are not already customers, or attracts a small segment of existing customers less profitably. Large service providers typically scoff at such proposals, but those initiatives might contain the seeds of technologies that transform the business and displace the present infrastructure. Marketers who recognize genuine disruptive technologies can influence the parent organization to invest and gain first-mover advantage. Marketers can also influence customers, especially pioneers and early adapters, to evaluate the new technologies and create ways to use them.

To their credit, several of the largest fixed-line telecommunications service providers invested in VoIP while it was still debated as an emerging technology. AT&T invested $1.4 billion in IP telephony provider Net2Phone in 2000. By early 2002, Atlantic-ACM already predicted 97 billion IP telephony minutes by 2007. In 2003, BellSouth partnered with Vonage to offer VoIP services over DSL facilities and focused other VoIP efforts on the small and medium-business market. In late 2003, Qwest undertook a replacement strategy, while SBC embarked on constructing a separate VoIP network. Complicating service provider decisions was uncertainty about whether VoIP would fall under the same regulatory umbrella as the remainder of the portfolio. Similarly, Verizon announced its migration to a packet-switched network. By 2004, any service provider without a VoIP plan began to risk displacement of its network by VoIP believers.

VoIP is a disruptive technology not only because it is cheaper and its performance lags that of the existing network. Its performance curve is steep compared to traditional voice services. Furthermore, its potential capabilities outweigh those of the existing voice network exponentially. After two years of availability in Japan, 10% of customers use VoIP services, even though users cannot receive inbound calls [3]. Thus some VoIP customers clearly are willing to use the technology in spite of its present limitations. While other factors differentiate the Japanese and American markets, such as cable penetration and the dramatic price advantage for VoIP calling in Japan (90% savings) over similar calls in the United States (50%), the direction is clear. Japan's experience demonstrates that VoIP services exhibit all of the price and quality characteristics of disruptive technologies.

But VoIP also offers enhancements due to its data capabilities, and this is where the disruptive technology might displace the 100-year-old telecommunications services model. Vonage uses a proprietary adapter between the standard telephone and the broadband connection. Ordinary calls appear to the customer as they normally do with a standard connection. Vonage goes beyond these capabilities because mostly everything about the experience is virtual. The system generates dialtone, which is meaningless without the central office (as it is in wireless). Ringing is similarly fabricated. Because the account is virtual, customers can choose a phone number anywhere Vonage serves. Furthermore, while traveling anywhere in the world, the customer can connect the adapter to a broadband connection and appear to callers to be in the home location of the assigned number. Calling between Vonage customers is free, and international rates lag most other long-distance plans. The customer can add numbers to ring to the same adapter for a very low price, gaining toll-free access to callers in another virtual area (assuming that toll-free access will continue to have value or even meaning). As another example, early VoIP services could not handle Enhanced 911. Vonage had its customers code their locations into the handset because it did not connect to the centralized 911 database. When Vonage does get its 911 working, though, it will have more functionality than the wireline 911 systems currently available. Integrating databases will enable paramedics to view the customer's medical requirements on a screen, for example [4]. As for the third litmus test, VoIP promises to cause incumbents to make the difficult decision to deploy it, though very reluctantly. Margins will remain low for VoIP services, and the incremental services that are possible will no longer achieve the same profits as equivalent enhanced services provided for incumbent service providers in the past. The steep performance trajectory will create a market with exponentially more services, an individualized interface, and price incentives to migrate more applications to the medium.

14.3 Maintaining an Innovative Organization

Pressures to innovate are strong in the telecommunications services industry. Advances in technology, industry consolidation creating well-funded players with an incentive to gain market share, deregulation, and an informed customer base contribute to the need for continuous innovation.

Disruptive technologies are most successful when their management is separate from the organizations supporting existing services. This reduces the conflicts between services with known markets and margins and their potential replacements that have neither markets nor brand strength. The size of the supporting organization should match the size of the known market, which means that disruptive technology groups should be small.

Innovation requires the support of an interested community, not only those assigned to its success. Some of these participants are outside of the organization, which partly explains their credibility to potential customers. Opinion leaders are those who exert informal influence over the behavior of others. They belong to customer trade associations, or they write for publications that customers trust, or they participate in user communities on the Internet. Change agents interact with the service provider and customer groups to resolve issues. Consultants interceding in the debates surrounding VoIP taxation fall into this category.

Inside the organization, those not assigned formally to the innovation's success can be valuable. Champions are most valuable when employees and customers see their initiatives as voluntary. Gatekeepers can be assigned or volunteers, and they foster innovation by bringing together resources that move the innovation forward. Continuous innovation requires the same senior management commitment, organizational flexibility, and focus that pervade all successful organizations. Innovation management is most effective when it respects but is not restrained by organizational boundaries, vertical or departmental, when it is iterative, tolerates uncertainty, and sets aside conventional wisdom.

BT Wholesale, the wholesale division of British Telecom, created the Innovation Unit in 2003. The unit's objective was to speed up time-to-market for applications. BT Wholesale announced three applications immediately, a mobile e-mail capability, interactive video for multi-participant chat rooms, and an interactive game.

Marketers can influence the success of innovative services by promoting the characteristics of new services that influence their diffusion into the marketplace. These attributes are relative advantage, compatibility, complexity, trialability, and observability. Relative advantage is the customer's belief that this new service is better than the one before it. Some breakthroughs in telecommunications technology improve service quality measurably in the laboratory without necessarily being observable by a particular customer. Latency in satellite communications is one example. No doubt many customers find call setup time to be an important differentiator, and many customers can differentiate between service providers. On the other hand, the service provider with the superior technology—no matter how insignificantly superior—is obligated to inform its customers that its technology is unsurpassed (even if it is nearly or arguably equaled) and that this attribute should be important to the customer. Compatibility means that customers view the new service as consistent with their existing values and needs. More to the point, in technology innovation, customers are reluctant to choose services that require investments in new hardware or are incompatible with the software and data upon which they already rely. Service providers that offer technology upgrades often minimize the disruption to customers by discounting any new hardware required significantly or bundling a payout invisibly in an annual contract. Of the five attributes influencing adoption, complexity is the only one that affects it inversely. While many early adopters are happy to study the intricacies of computer code, innovations do not convince the mass market needed for success until its complexity is minimal. Trialability addresses the customer's reluctance to commit to the innovation without experiencing its benefits first-hand. Free service can resolve this problem for monthly subscribers, and aggressive service providers will allow qualified leads to test-drive new hardware before buying. Observability enables the customer to see the benefits before deciding to buy. For telecommunications services, free trials assist in gaining this advantage.

Service providers need to evaluate their own strengths and core competencies and the weaknesses and strengths of their competitors. By definition, the service provider's strength is a unique capability, and its uniqueness prevents competitors from retaliating. The most enduring innovations within the telecommunications services sector had strengths that made them impervious to competitive retaliation. Examples such as MCI's Friends and Family, AT&T's wireless Digital One Rate program, AOL's instant messaging, and Nextel's push-to-talk were examples of such innovation. In each case, the service provider used its greatest strength as the cornerstone of its innovation. MCI's information infrastructure was new and the best of its time, and the long-distance provider could rate calls using a complex algorithm with relational databases. AT&T, with its legacy systems already burdened with most of the market, could not hope to retaliate. AT&T had more success in the wireless market, where it could compete against the incumbent local service providers with its own long-distance network. At the time, the market was a duopoly—only two licenses granted in each market—and AT&T's likeliest competitor by far was a regional Bell operating company that was precluded by regulation from owning long-distance facilities. AOL's instant messaging used the network effect to enhance the value of instant messaging. By restricting its user base to its own customers—representing most of the market that was interested in messaging—the Internet service provider created a favored destination for new customers who wanted instant messaging. Its competitive advantage was so strong that AOL attracted considerable attention from regulators. Nextel appealed to the business customer's buying criteria that differed from those of churn-prone consumers. Nextel's customers sought value, including a costly handset that lowered operating expenses considerably, and a tool that fostered timely communication between employees and with customers.

Some innovations preclude retaliation not only because of their competitors' weaknesses, but from their competitors' strengths as well. MCI and Nextel also leveraged the strengths of the industry leaders as weaknesses that were not only impossible but also undesirable to imitate. In the case of MCI's Friends and Family, AT&T could not hope to provide deep long-distance discounts between pairs of its own customers, as they represented nearly the entire calling profile of the market. Nextel's competitors were fighting for the huge consumer market, where a low-cost handset was very important and the ability to churn was a benefit.

In another example of a competitor's strength creating an opportunity from competitive weakness, Vonage's innovative business model has capitalized on the glut of intercity facilities and the ubiquity of broadband access outside of the conventional telephone network. Its independence regarding network facilities enables it to seek the lowest prices without favoring its own network.

Moreover, independence from reselling the existing access facilities of fixed-line telecommunications service providers shields Vonage from the potential influence or even obstruction from incumbents, who appear to be in no hurry for packet-switched networks to gain critical mass. Its unbranded and branded VoIP offers capitalize on the incumbent fixed-line providers' strength—large investment in transmission and distribution assets—to prevent them from seeking similarly flexible partnerships. This flexibility also affords Vonage time to hone its service offerings and create the marketplace in the image of its own strengths. This affords first-mover advantage and sets the customer's expectations to its own service parameters.

14.4 Case Study: NTT DoCoMo's i-Mode

In its first year of operation, NTT DoCoMo's i-Mode data service gained 4 million subscribers. By 2004, more than 40 million of its 46 million mobile customers had opted for the platform, and its churn rate was half the rate of its voice-only wireless customers. Its success was in stark contrast to the failure of the similar service WAP to take hold in Europe (a Google search on the point turned up English, French, German, and Dutch headlines that WAP was a FLOP). What did Japan's NTT DoCoMo do to entice subscribers to its innovation?

The wireless service provider focused on content, not technology, viewing the wireless solution as its customers would, on outcomes. While WAP advertisements crooned about the Internet, GPRS, and WAP, i-Mode promoted "fun." An affordable bundle of services included e-mail and some browsing for basic information, but much of the premium information services were outsourced to partners. NTT DoCoMo avoided referring to Internet access so as to manage the expectations of customers. Though the service originally targeted business customers, the eventual customer base was disproportionately made up of consumers, especially the youth market.

Outsourcing enabled NTT DoCoMo to focus on its own strengths in operations and gain the benefits of brands their customers already regarded highly. The vast majority of the premium services revenues went straight to the content providers, giving them an incentive to develop more content independently. In contrast, WAP service providers had developed much of the content on their own. i-Mode enabled NTT DoCoMo to develop an emotional connection to the customer and benefit from branded English-language sites such as Dow Jones, CNN, Bloomberg, and Disney as well as popular Japanese content providers. Customers could manage financial accounts, pursue dining and entertainment, shop, and seek information from the content providers they trust. Content partners adhered to specific guidelines concerning the quality and timeliness of the content they provided.

Another incentive for content providers was the billing arrangement. DoCoMo provided centralized billing, making it worthwhile for content providers to offer low-priced content that would not cost justify if they needed to bill for it on their own. NTT DoCoMo's 9% commission left most of the revenues to those who supplied content. Moreover, uncollectibles for such small amounts were less likely in the billing clearinghouse; the customer whose account was overdue risked losing all i-Mode subscriptions, features, and services. The billing application provided users with unique capabilities, for example, to view their daily or monthly spending in text or bar chart format.

Partnerships and the use of open standards were part of i-Mode's success formula. NTT DoCoMo communicated its device requirements to handset manufacturers and exerted control over the handset specifications. The handset device had a color screen, in contrast to WAP applications in monochrome. The service provider developed partnerships with many of its suppliers, including Hewlett-Packard, Japan Air Lines, Macromedia, Microsoft, Oracle, Samsung, Sony, Sun Microsystems, and VISA. Supported by contracts, these partnerships gave the supply-chain partners a stake in i-Mode's success.

All of these decisions: a content focus, outsourcing, partnerships, and flexible billing, gave i-Mode a reduced time-to-market and a valuable first-mover advantage, able to withstand similar competitor entry within months of its launch. NTT DoCoMo leveraged the capabilities of outsiders by the offer of high commissions providing powerful incentives to participate. The service provider similarly provided customers with effective incentives to take monthly subscriptions to information services, intentionally priced at the cost of a magazine. A network effect followed its early growth: customers could e-mail other customers, and content providers could reach more subscribers. i-Mode's lead actually accelerated once competitors entered the marketplace.

Customers appreciated the per-packet pricing, which, though usage-sensitive, was low compared to alternative ways of performing the same tasks and avoided per-minute concerns about data access rates that tended to hamper early WAP usage. A baseline amount of free packets insulated the customer from charges for potential spam. Content providers charged very low fixed monthly fees for premium services, creating simple, easy-to-bill, and easy-to-promote pricing. The always-on service meant that customers did not need to log on and off whenever they wanted to connect, but they only paid when they downloaded information. WAP pricing suffered from per-minute charges, but GPRS services solved that pricing problem with the capability of charging per packet. Average i-Mode monthly usage-based charges to customers were similar to wireless plans offered with much lower functionality in other countries. In 2003, DoCoMo responded to competitive initiatives and the growing monthly usage of its customers by offering a flat-rate alternative to usage-based packet pricing. Customers could also obtain discounts by prepaying annually.

The use of iHTML, (also described as c-HTML or compact HTML) a subset of Internet standard HTML, as the markup language enabled site owners without contracts with NTT DoCoMo or the desire for compensation to modify their sites and offer unofficial access to customers through the phone. Users simply typed in the address of the site on the keypad to reach tens of thousands of the unofficial sites.

Within partnerships with local wireless service providers, NTT DoCoMo expanded its i-Mode service to Germany, the Netherlands, France, Spain, Italy, Belgium, and Taiwan and was in negotiations with AT&T Wireless at the time of the merger with Cingular in the United States. Expansion tests any business model, and i-Mode's growth faced challenges in all of its new markets. Technology was one issue. Japan was a world leader in establishing wireless technologies, including broadband. NTT's partners faced investing in new generations of network technology and educating a customer community outside of Japan that was already skeptical about wireless Internet because of their dissatisfaction with WAP. Another challenge was market structure. While NTT DoCoMo already owned more than half the Japanese market at the time of the i-Mode launch, other markets are less concentrated. For example, even if it partnered with the market leader in the United States, NTT DoCoMo could offer only about one-quarter of the market. The service provider had bet on AT&T Wireless, with less market share than most U.S. service providers and an apparent unwillingness to invest in broadband technology, but it was rescued by AT&T's absorption into a stronger competitor. i-Mode's menu-driven interface and text-based communication fit with Japanese language and culture, but did not offer competitive advantage in the West. A third challenge involved embedded technologies. i-Mode offered Japanese consumers Internet access and transactions when few had alternative access channels, but U.S. consumers were on-line, shopping, browsing, and conducting transactions from the comfort and wide screens of their own PCs. To achieve worldwide success in wireless broadband, NTT DoCoMo would need to adjust its model overseas.

References

[1]   Drucker, Peter F., *The Practice of Management*, New York: Harper and Row, 1954.

[2]   Christensen, Clayton M., *The Innovator's Dilemma*, Cambridge, MA: Harvard Business School Press, 1997.

[3]   "Japan's VoIP Experience Could Preview Things to Come in America," *Telecom Policy Report*, Vol. 1, No. 25, 2003.

[4]   Jackson, Donny, "VoIP Recognition," *Telephony*, Vol. 244, No. 2, 2004.

15

Customer-Centered Technology Marketing

15.1 The Uniqueness of Technology Services Marketing

Much of the marketing for traditional telecommunications services resembles the selling of commodities, with known products, known benefits, and clear customer beliefs concerning value. On the other hand, technologies new to the buyer—especially those for which marketing is crucial—are uncertain as to their capabilities, their worth, and even their application. The first-time buyer of broadband to the home, or broadband wireless, or IP-based services, stands somewhere on the diffusion of innovation curve described in Chapter 1.

Goods (in the economic sense) can be search goods, experience goods, or credence goods. Search goods include those products and services about which customers can learn enough simply by conducting sufficient research. In the telecommunications services market, one example of search goods might be discount long-distance service. Many customers are willing to assume that the quality of all competitors is similar and that all are acceptable. These services respond well to informative advertising, as information helps customers make purchase decisions. First-time buyers of new technologies are unlikely to purchase solely on the basis of information search.

Other telecommunications services are experience goods, that is, the customer does not know their value until the service has already been consumed. One example is broadband access for the first-time buyer. Customers do not know in advance of their purchases how the faster service will affect either their usage or their productivity. Advertising plays a role in selling experience goods, but an indirect one, in favor of promotions. Betting that customers will love what they see, broadband service providers are often willing to offer incentives in the form of hardware adapters for fixed-line customers or free temporary usage for wireless users. Experience goods can present switching costs to the customer in the form of uncertainty. The customer that is satisfied with the network quality of a telecommunications service but dissatisfied with customer service, for example, might be reluctant to switch to an alternate provider. Outside of other considerations that the customer can compare, such as price, this customer cannot know whether the competitor's network quality or even its customer service is an improvement over the present provider's.

Lastly, some goods are credence goods. This peculiar sort of service is so difficult to judge that customers do not know whether the brand they selected is best even after they have purchased it. Some telecommunications services fall into the search goods category at the low end and the credence goods category at the high end. When a customer selects a long-distance provider because a consumer group rates its quality the best, that customer has treated the service as a credence good. Most fixed-line customers cannot hear the small differences in service quality that might affect dial-up modem speed by a few thousand bits per second. Most wireless customers do not even use the service in enough territory to judge coverage against competitors, might never have used a competing service, or might have used a competitor in the past but never held a contest to test service quality at the same time and in the same spot. Nonetheless, business customers read trade publications, visit conferences, converse in user groups, and listen to salespeople. Consumers hear the news, talk to friends, and watch commercials. Credence goods are about reputation, or more simply, the brand. Positioning a service to have credence good characteristics makes the service less vulnerable to competitive attack, but its premium becomes vulnerable instead to self-destruction through unwise product extensions, corporate scandal, or external factors that might be unrelated to service quality. Furthermore, customers who gain their knowledge about service quality only through company-sponsored advertising might become justifiably skeptical of its validity.

Most telecommunications markets are characterized by significant supply-side economies of scale. These economies have their most significant impact on new services. Though the creation of the first broadband minute requires enormous investment; the second minute is almost free by comparison. This creates pricing opportunities and unusual strategies to gain market share. Providing a wireless handset at no charge or a deep discount in exchange for an annual contract is an enduring strategy for wireless service providers. Pricing magazine subscriptions to compete with paper alternatives instead of basing prices on fully allocated costs gave i-Mode a huge customer base and high incremental margins.

Network externalities, also called the network effect, contribute to the actual and perceived value of competing technologies (bearing in mind that most of the actual value of a technology is the customer's perception anyway). Network effects exist when the value of a good, or more specifically the demand for the good, rises as the number of customers increases. As a matter of fact, economists refer to this situation as demand-side economies of scale. Historically, the example economists used to describe the network effect was, in fact, telephones. A century of convoluted pricing could survive because businesses were happy to overpay for service if all consumers would just get telephones and call them. Network effects can be direct or indirect [1]. Direct network effects increase the value of a service directly when more customers exist, like the telephone network. E-mail involves a direct network effect; the more customers with e-mail addresses give new e-mail customers more reason to be on-line. Nextel's push-to-talk and Verizon's IN-network plan are other examples; the more customers each service has, the larger and more valuable the universe of potential calls. Indirect network effects increase the value of the service indirectly. Broadband penetration enables users to download content-rich pages, and that causes Web site developers to enrich their sites.

Many technology markets also include switching costs and lock-in, though regulators properly do whatever they can to reduce those costs, and competitors do whatever they can to make them irrelevant. Local number portability ceased to constitute a cost of switching in most markets, but others remain. Bundling services using deep discounts can be an effective lock-in technique. E-mail addresses continue to constitute the best lock-in that otherwise indistinguishable ISPs hold, sparking significant antispam activity and a host of Web-based e-mail providers who offer Web portals with ancillary services in addition to independence from those who offer Internet access alone. NTT DoCoMo's i-Mode created a lock-in with its customers by requiring an upfront investment in the hardware. At the same time, the wireless service provider reduced switching costs for its suppliers, the content providers, to join i-Mode by limiting the amount of investment required to convert their existing HTML code. Wireless service providers continue to insist on annual contracts with substantial penalties for early termination, which create a somewhat artificial lock-in for customers. Nonetheless, providing hardware as a bonus for a contract might serve to reduce the customer's perception of switching costs. The purchase of a modem or handset or adapter for hundreds of dollars at the inception of a no-commitment subscription might serve as more of a barrier to end-of-year churn than any contract would have done. Consumers are much more sensitive to upfront costs than ongoing expenses. Open standards tend to reduce lock-in, so market leaders tend to resist them, though maintaining proprietary standards might not be wise in technology markets. Opening standards to third-party developers can create a complementary market and indirect network effects.

Technology marketing also experiences a situation called "the technology paradox." Prices fall so fast and so low in technology markets that many developers cannot hope to recoup their initial investment. Moreover, technology service providers find that achieving a critical mass of customers is much more important than recovering research and development costs. Teleport Communications, a competitive local service provider later acquired by AT&T, installed a dozen optical fibers—a million times the customer's needed capacity—for its customers because the incremental cost of installation (marginal cost, to economists) was tiny and offered great profit opportunities at a later time [2]. Even before deregulation and the attraction of fax and data, monopoly local telephone companies historically installed two lines for each one ordered in each home, hedging the possibility of additional lines or the failure of the original installation. Verizon's IN-Network program offered the most valuable free minutes to customers on the market, exploiting its market leadership. Verizon's expectation was that customers seeking in-network minutes would cover those incremental costs with the out-of-network minutes they purchased. Furthermore, the service provider realized that market leadership on the brink of broadband adoption would be a valuable asset.

Another characteristic is of technology markets is the concept of "free-riding." Free-riding occurs when some customers get pirated products or services—such as software, music, or satellite television—while paying customers foot the bill. It is a special problem in technology markets because the incremental cost of the next customer approaches zero. While it seems as though no harm is done to a network by allowing a handful of happy pirates, most free-riding is undesirable to content providers and service providers because it reduces the incentive and consequently the resources to create new content and services. Free-riding does have a benefit, though, and that is market share. Media mogul Rupert Murdoch has called the pirates stealing News Corp.'s Star TV satellite signal "splendid entrepreneurs," because they increased the market distribution and enabled News Corp. to raise its advertising rates.

Technology vendors utilize versioning to make use of the low incremental cost of additional customers. Versioning refers to the practice of selling slightly different forms of a product or service to meet the distinct needs of users. Service providers who offer DSL in various speeds are using versioning to expand their markets. The challenge is to set up the versions so that customers self-select the service that best meets their needs and that provides maximum profit for the service provider. DSL versioning segments customers against the speed of service. Required on-line billing on discount subscription accounts segments customers by channel. Offering an advertising-free e-mail interface, a popup-free browser experience, premium content on a customer service portal, or portions rather than complete content all represent versioning opportunities for service providers. For example, a wireless service provider could download instant stock quotes for several selected securities at no charge to its customers with the highest lifetime value instead of the 20-minute delayed quotes readily available on most portals. Versioning offers an opportunity to practice price discrimination at the customer's request based on their willingness to pay. Because versioning at the low end can expand markets beyond what they would be without versioning, this practice can intensify the network effect, drawing more customers and increasing profitability. Thus lowering some prices might have the effect of creating market leadership.

Cisco's Ron Ricci and John Volkmann of Advanced Micro Devices set out to deconstruct the momentum of technology brands [3]. After surveying 20,000 consumers and business buyers of fast-changing products, they borrowed from physics to identify six forces within the three categories of mass, speed, and direction, calling them "forces of differentiation."

Within the category of mass, the forces include the relevance of the value proposition, its ecosystem potential, and its category leadership. Value proposition relevance refers to the importance of the brand to the success of its customer. Here, digital products depart from nondigital brands. Customers expect nondigital services to perform to the same specification time and time again. Customers judge digital services on their present performance but also on their future: will this service compete well in the future? When can I upgrade? Will the upgrade be worthwhile? Its ecosystem potential refers to its importance with its category partners to the success of customers. Here, the partnerships between service providers are crucial. For example, local service providers have created partnerships to offer service bundles to customers. The choice of service partner can derail the service provider's carefully nurtured brand. Its category leadership refers to the product's dominance within its category and the category's importance to the customer. Customers are attracted to market leaders, whether or not network effects exist. Market leaders also inspire complementary products and third-party support. While most telecommunications services remain undifferentiated, complementary products and services will drive purchasing in software-driven telecommunications services markets of the future.

The brand speed category assesses how well the brand is able to manage change in its marketplace. This includes anticipating the demand for new technologies and pacing its own technology introductions to optimize the adoption cycle. The brand direction category includes both vision and integrity. Vision refers to the brand's articulation of its future direction by credible leaders that act as evangelists for the imagination. Integrity refers to the competence of the brand, its reputation, and the customer's belief that the brand lives up to its promises. The authors conclude that analyzing the brand's attributes is only the beginning. The next step is to conduct a similar analysis of competitors to find opportunities for differentiation. Thus armed, the service provider can focus on the points of positive differentiation and attack competitors along those attributes.

15.2 Revisiting the Diffusion of Innovation Curve

The diffusion of adoption curve summarized in Chapter 1 provides insight beyond simply labeling groups by their rate of adoption of discontinuous or disruptive services. Service providers need to adjust their marketing strategies and tactics to concentrate on any point along the curve to maximize the likelihood of success. At each transformation point on the curve, there is a crack defining the different buying criteria of customers. The crack obstructing the migration from early adopters to mainstream critical mass is so large that consultant Geoffrey Moore calls it "the chasm" [4]. Most high-technology failures occur because the technology cannot cross this chasm into mainstream markets. This event and its ramifications are superimposed on the diffusion of innovation curve in Figure 15.1.

Consultant Moore maintains that early adopters are more willing than the early majority to change their behavior to adjust to new technologies, but mainstream customers will wait until the technology is easy to use. The chasm occurs between the early adopters and the early majority. Early adopters are not technology innovators, but they are able to imagine the benefits they would gain from the new technology and make their purchase decisions without requiring outside recommendations. Marketers need not focus on early adopters, who prefer to develop their own opinions. In fact, marketers are wise to ignore stated customer desires in this phase. The early majority, on the other hand, is not reluctant to accept new technology, but they need outside confirmation before making the investment. They look to the early adopters for guidance.

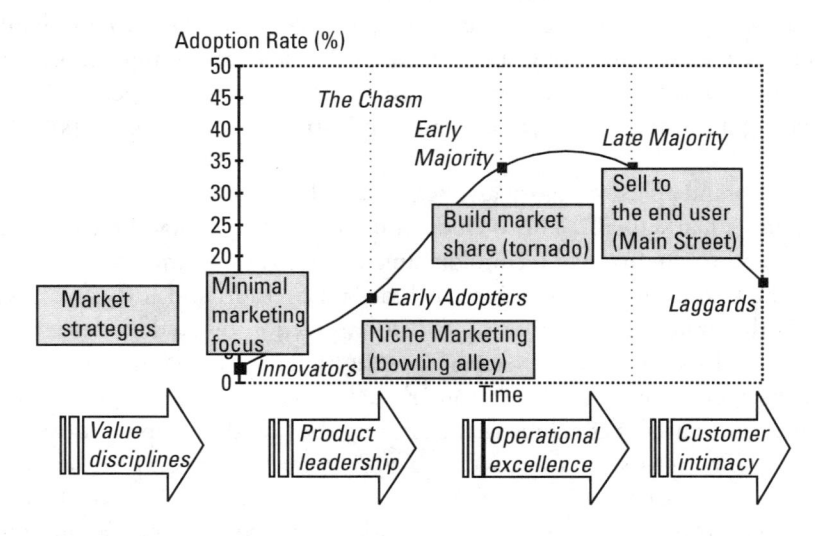

Figure 15.1 Revisiting the diffusion of innovation curve. (*After:* [4].)

The chasm describes the difficulty of capturing the early majority, which makes up a third of all customers, and is a vital element in the success of any new technology. This view implies that the marketing of disruptive technologies requires multiple service introduction strategies, with a new approach for each adoption segment. While the early adopters are technology enthusiasts, the early majority consists of pragmatists. These customers, unlike their visionary predecessors, require a complete, functioning service that is easy to set up and easy to use. For customizable technology services, serving this entire customer group—while necessary for eventual market leadership and success—is impractical. Moore recommends that the service provider select a subset of the group, such as a vertical market or a clearly defined customer segment, and focus market development on producing a compelling reason to buy.

This period, which Moore likens to a bowling alley for its tunnel-like focus, should concentrate on differentiating the service, emphasizing value and return on investment to customers, and assigning a low priority to competition or infrastructure. From success in one market niche, the service provider can then use its credibility in one market to address new niches, and eventually develop the brand leadership that convinces the early majority to buy. During this phase, customers in multiple segments rush in like a tornado, and the marketing strategy focuses on distribution, infrastructure, and competitive pricing as the service provider builds market share. Finally, the mass market begins to buy, and the strategy changes again. Prices begin to erode, because new customers are arriving at a slower rate.

The marketing strategies along the adoption curve for disruptive technologies intersect the value disciplines of market leaders discussed in Chapter 3 [5]. Each of the value disciplines, taken alone, is an appropriate strategy at some point on the diffusion of innovation curve. During the earliest phases, product leadership creates disruptive change for the innovators and early adopters. These customers seek performance and will buy an unproven technology. They are comfortable with risk and do not consider reliability or ease of use to determine whether to purchase. A strategy of technology leadership is most effective with this group.

As the market builds to greet the chasm, the service provider needs to adjust its market strategy from product leadership to operational excellence. Early and late-majority customers demand wide distribution, ease of use, and reliable performance. These are the characteristics of a service provider whose core competence is operational excellence. Service providers redirect their investments away from research and development and towards the infrastructure that will ensure supply, minimize returns, and provide support.

Finally, customers who adopt new technologies in the late majority and laggards who resist adoption for as long as possible respond best to customer intimacy. Long-standing marketing wisdom and a matrix model from Boston

Consulting Group viewed this period for harvesting, drawing profits from "cash cows" without making further investments. This famed growth/share matrix defined all business units in terms of their position against market share and market growth. High-growth businesses with a large market share are "stars" and require investment during the growth phase. These units represent the future profitability of the enterprise. Low-growth and low-share units are "dogs" whose resources should probably be redirected elsewhere. High-growth products with a small market share are "unknown," for which investment might or might not result in creating new stars. Last, high market share businesses in mature markets of low growth are cash cows. These cash cows are evident when the laggards and the late majority finally come around to adopting new technologies. Boston Consulting Group recommends that mature offerings constitute only a part of the service provider's portfolio. Profits from cash cows should be available to fund new stars, but disruptive technologies tend to supplant sustaining technologies before they spend much time in harvest.

For business customers, adopting new technologies gives rise to another challenge for service providers and other suppliers. This challenge, called the assimilation gap, occurs when an organization acquires a technology but fails to adopt it wholeheartedly [6]. Figure 15.2 depicts this gap against the technology S-curve.

The first S-curve demonstrates the acquisition of the technology, which in this case is the point that all hardware and software has been installed, users are trained, and the technology is ready to use. The second S-curve represents the lag after its purchase and installation before users actually incorporate the technology into their business processes. The delay between the installation of a new service and its deployment by users represents the gap. The importance of this gap is exacerbated by the conflict between increasing returns to adoption—that is, the benefit of network effects as the network of users increases—and knowledge barriers impeding adoption, in other words, the difficulty to learn how to use the technology.

The importance of this gap to telecommunications service providers will grow as technologies become more complex and more software intense. Most of the research concerning the assimilation gap currently centers on software development, but IP-based telecommunications services customized to mission-critical business telecommunications environments will likely operate in a similar manner. Service providers will need to ensure that this lag is as short as possible. First, users that become familiar with the technology faster will realize its benefits sooner. Second, customers that are well trained will use the technology more effectively and will achieve higher benefits. Finally, minimizing the assimilation gap will increase the lock-in that keeps customers from abandoning the technology in favor of competitors.

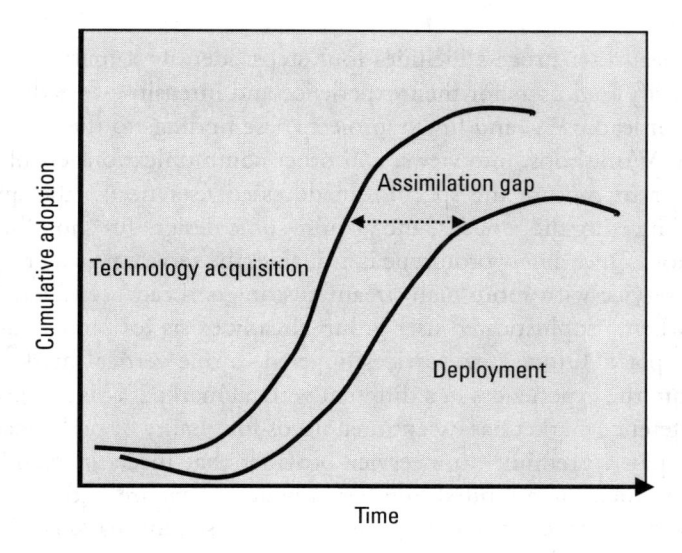

Figure 15.2 The assimilation gap.

15.3 Technology Marketing Tools

One challenge to developing and selling technology services innovations is the inability of conventional market research tools to assess customer demand accurately. Though much of the research surrounding technology development has focused on products, software-based services design will be available to service provider marketers to meet the needs of targeted segments. Potential customers simply do not have adequate experience with the new services to give market researchers legitimate reactions. Among the best-known tools to improve market research are lead users, empathic design, and quality function deployment. These market research techniques depend less on what customers say and more upon what they do.

MIT professor Eric von Hippel developed the Lead Users technique as one approach to defining the market for new technologies [7]. Lead users are those whose unusual present needs will represent those of the mainstream in the future. They correspond roughly with the innovators on the diffusion of adoption curve, but differ in that their strong immediate functional needs eclipse the typical innovator's desire simply to maintain a leading technology edge. Lead users are also highly motivated to innovate because a solution will benefit them significantly. Though they do not develop technology for its own sake, as do technology innovators on the adoption curve, lead users often cobble together their own prototype solutions because they are not yet available in the marketplace. Lead user analysis identifies these atypical but valuable users of mainstream services.

The lead user process includes four steps: identify a market or technical trend; identify lead users for their experience and intensity of need; analyze the need data of lead users, and finally project these findings to the general market of interest. Workshops, interviews, and other communications enable the service provider to capture the specific unaddressed user needs, the quantifiable benefits of meeting those needs, and acquire some design direction for the eventual solution. Once a new prototype is available, the service provider can test the proposed service with more mainstream customers. Lead user analysis is most effective when a sophisticated user group already exists for services available in the marketplace. Often, new services targeted to one vertical market can gain insight from the experiences in a different vertical market. This is especially true when a reference market has exceptional needs for quality or performance and is willing to pay a premium to a service provider that meets its standards. The research can focus on a limited number of benefits that are critical to the reference market and have a potential position in mainstream markets, while ignoring other attributes that are less relevant to mainstream application.

Telecommunications equipment manufacturers have turned to lead users for inspiration in developing commercial applications. For example, Nortel launched a lead user initiative to study Internet-enabled wireless applications, especially location-based services. Nortel's reference users included representation from fields as disparate as military battle management, remote diagnostic field technicians, mobile telemedicine, law enforcement, aviation specialists, oil field operations, and remote news broadcast operations. The equipment manufacturer later included animal trackers and storm chasers. All of these lead users, disparate as their areas of endeavor might be, share two characteristics: they need highly targeted innovation before most of the universe of wireless customers, and the performance of new technologies within their fields provides substantial benefits to them.

Empathic design observes the customer in the context of using the service, to capture the attitudes, behavior, and unmet needs that the customer cannot articulate [8]. The majority of customers will not ask for services they do not know are technically possible. Customer reports of their own behavior and even their attitudes can be unreliable. Observers might notice that the customer uses a workaround when an existing feature would address the need better. The observation can record verbal and nonverbal activities and sensory or emotional reactions to service attributes. The researcher can uncover the triggers of use, the ways the customer uses the service in one's own environment rather than in a laboratory, the ways each customer has tailored the service to unique needs, intangible aspects of the service, and unarticulated needs. For example, watching a customer using a wireless device might disclose that the business customer uses the phonebook much more often than the spreadsheet or vice versa, or that the customer uses the camera for capturing competitor street advertising, or that the customer has labeled the case of the telephone with the telephone number. These observations might reveal opportunities for highly targeted new services or upgrades to the device itself. The empathic design process begins with observation of the customer using the service in real settings, even if that means getting permission to follow a customer around for days or sit by the customer for hours at the office or at home. The next step is to document the customer's behavior and problems they face, including the customer's responses to open-ended questions such as "why are you doing that?" Then the observation teams pool their experiences and brainstorm solutions to the problems and opportunities they discovered. The last step is to develop prototypes of potential solutions.

Empathic design is especially useful for disruptive technology research, because it refrains from asking customers questions that they simply cannot answer about whether this new technology will help them. Furthermore, empathic design concepts apply well in the virtual environment, where Internet domain owners can examine user behavior as it occurs. Service providers offering on-line billing or service descriptions can track what errors customers commonly make, which pages customers visit most often, where they came from, and where they go next. America Online changed its innovation focus from information content to communications services between subscribers after monitoring customer-usage patterns.

Quality function deployment (QFD) grew out of Japanese management and engineering methodologies. Its goal is to include "the voice of the customer" in service development. QFD proposes a phased development plan that involves cross-functional teams of marketers, engineers, customers, and others in all stages of development. QFD proposes that though customers are not necessarily experts in product/service design or operations, they know better than anyone else how they use the services and what benefits they derive. The discipline attempts to quantify user needs and then prioritize them according to their importance to the customer. QFD is not satisfied when the customer says that fast response time is important; the customer needs to verify that "fast" means one second, or four seconds, or milliseconds. Customers also compare the service against competing services, which provides additional insight. The next phase deconstructs those needs into the elements of design that can create the desired outcome. For example, if a customer said that response time of one second was the most important criterion in selecting a service provider, the next phase would examine the elements that contribute to response time and conduct research to locate best-in-class processes. Significant emphasis on what to design occurs before the actual design process takes place. The process examines attributes to determine the customer impact of improvements by categorizing them into specific types of customer needs. If an increase in the quality has a linear impact on customer satisfaction (such as additional bundled minutes), it meets a performance need. If reductions in quality would have a substantial negative impact (such as service quality or wireless coverage), it meets a basic need. If a new attribute might make an exponential change in customer delight, it is an excitement need and is a probable source of competitive advantage.

The design phase approaches the deconstructed needs, design criteria, and competitive data as a "house of quality," which is a design matrix that combines the customer requirements (voice of the customer), design parameters (the voice of the engineer), and competitive positioning data. Like most marketing processes, these customer-centered research tools are supported by software packages.

In summary, customer-centered marketing requires the service provider to focus on solutions rather than features, demand instead of supply, and customer behavior rather than attitudes. Most service providers—and all effective service providers—are advantaged by the fall of regulation, the transparency of the market, growth in demand for higher volume services, advances in technology, and customer engagement in the purchase process. Nonetheless, advantage does not always equal success. The service providers that become market leaders will do so by winning one customer at a time, again and again, throughout the customer relationship lifetime.

References

[1] Katz, Michael L., and Carl Shapiro, "Network Externalities, Competition, and Compatibility," *American Economic Review*, Vol. 75, 1985, pp. 424–440.

[2] Gross, Neil, and Peter Coy, "The Technology Paradox: How Companies Can Thrive as Prices Dive," *Business Week*, No. 3414, 1995, pp. 76–82.

[3] Ricci, Ron, and John Volkmann, *Momentum: How Companies Become Unstoppable Market Forces*, Cambridge, MA: Harvard Business School Press, 2003.

[4] Moore, Geoffrey A., *Crossing the Chasm: Marketing and Selling High-Tech Goods to Mainstream Customers*, New York: HarperBusiness, 1991, 1998.

[5] Treacy, Michael, and Fred D. Wiersema, *The Discipline of Market Leaders: Choose Your Customers, Narrow Your Focus, Dominate Your Market*, Reading, MA: Addison Wesley Longman, 1995.

[6] Fichman, Robert G., and Chris F. Kemerer, "The Illusory Diffusion of Innovation: An Examination of Assimilation Gaps," *Information Systems Research*, Vol. 10, No. 3, 1999, pp. 255–275

[7] von Hippel, Eric, "Lead Users: A Source of Novel Product Concepts," *Management Science*, Vol. 32, No. 7, 1986, pp. 791–805.

[8] Leonard, Dorothy, and Jeffrey F. Rayport, "Spark Innovation Through Empathic Design," *Harvard Business Review*, Vol. 75, No. 6, 1997, pp. 102–113.

Acronyms

ARPU average revenue per user
ASP application service provider
CART classification and regression tree
CHAID Chi-Square Automated Interaction Detection
CRM customer relationship management
DSL digital subscriber line
ERM enterprise resource management
FCC Federal Communications Commission
GPRS general packet radio service
HTML Hypertext Markup Language
ISP Internet service provider
PRM partner relationship management
SCM supply chain management
SMS short message service
SSM soft systems methodology
VoIP voice over Internet protocol
WAP wireless application protocol
Wi-Fi wireless fidelity

About the Author

Karen G. Strouse is the owner of Management Solutions, a consulting firm specializing in the changing telecommunications industry. Her client list includes large and mid-sized service providers, and small telecommunications companies serving targeted industry segments.

Ms. Strouse has assisted telecommunications service providers in developing entry strategies for new lines of business, developed and led seminars for strategic planners and marketing managers, and conducted industry and competitive analysis in support of diversification efforts. She has assisted management in devising strategies, marketing approaches, pricing, and infrastructure to support new ventures. Ms. Strouse has also assisted clients in preparing business and strategic plans, developing acquisition strategy, performing competitive analysis, and improving operational effectiveness. In various industries, she has directed teams to improve business operations through reengineering of business processes, and assisted management in constructing effective and efficient organizational structures.

Her experience includes a 13-year career with Bell Atlantic Corporation (now Verizon), serving in a variety of assignments, including director–strategic planning. After serving as a consulting manager with international firm Deloitte & Touche, Ms. Strouse founded Management Solutions in 1992. Artech House published her first book, *Marketing Telecommunications Services: New Approaches for a Changing Environment*, in 1999, and her second book, *Strategies for Success in the New Telecommunications Marketplace*, in 2000. Her e-mail address is kstrouse@yahoo.com.

Ms. Strouse holds a B.A. and an M.A. in communications from Temple University and an M.B.A. in finance from St. Joseph's University in Philadelphia.

Index

Advertising, 158
America Online (AOL), 69, 70, 118, 179
Application Service Providers (ASPs), 82
Appreciation programs, 137
Assimilation gap, 191
AT&T, 34, 127, 172
    brand equity, 160
    Digital One Rate, 25, 70, 179
    low-cost dial-around service, 21
    Wireless, 127, 182
    as wireless service provider, 23
Attentiveness, 136
Average revenue per user (ARPU), 142

Bandwidth, as commodity, 165
Barrier to entry, 17–18
Behavior segmentation, 106–7
Benefit segmentation, 108
Billing
    on-line, 85, 148
    systems, 147–50
Brand equity, 160–62
    defined, 160
    high, 160
    managing, 161–62
    strong, 160
    tests, 161
    tracking, 161
Branding, 118, 157–69
    commodities, 164–66
    product positioning and, 162–64
    rebranding, 160
Brand(s)
    advertising and, 158
    best-known, 166–67
    company leadership and, 159–60
    customer care and, 159
    direction category, 187
    identity, discarding, 167
    image conveyance, 159
    importance, 157–60
    leadership, 168
    line extension, 168
    management, 162
    marketing-controlled management, 158
    nurturing, 160
    press and, 160
    protecting, 166–69
    restructuring, 167
    service quality and, 158
    speed category, 187
    strength, measurement, 168
    success, 161
    transforming commodities to, 166
Bundling, 123–25
    benefits, 124
Bundling (continued)
    cable company, 125

churn and, 136
defined, 123
discounts, 124
innovation and, 25
local service provider, 124–25
successful, 24–25
video services, 125
Business buyers, 8, 9, 10
cost effectiveness and, 9
decision-makers as, 8, 9, 10
knowledge, 9
as knowledgeable, 19
service provider relationship, 10

Call-detail analysis, 149
Capital-intensive industry demand, 51–54
defined, 52–53
estimating, 53
increasing, 54
See also Demand
Channel management, 77–89
cross-channel strategies, 81–83
e-business, 84–86
evolving routes to market, 78–81
PRM, 83–84
scoping the channel, 87–89
strategic alliances, 86–87
strategies, 87–89
traditional distribution channels, 77–78
See also Distribution channels
Chi-Square Automated Interaction Detection (CHAID) analysis, 110
Churn, 131–34
benefiting, 140–42
bundling and, 136
consumer likelihood to, 133
costs, 132
costs, limiting, 137–40
defined, 131
double-digit, 110
forms, 132
intention, testing, 135
intuitive response, 132
limiting, 131–34
management process, 140
predicting, 131–34
rates, 133
as self-fulfilling prophecy, 142
for wireless discounts, 134
See also Loyalty

Cingular, 34, 169
Classification and Regression Tree (CART) networks, 110
Cluster analysis, 110–11
forms, 110
phases, 110
uses, 110
Commodities
bandwidth, 165
branding, 164–66
defined, 20, 164
long distance as, 165
marketing services as, 20–22
pricing, 117–18
Competition, 15–25
intensity of rivalry, 16
local telecommunications services, 52
supplier pressure and, 18
Competitive intelligence, 63–75
challenges, 65
defined, 63
developing, 63–68
goals, 67
parity and, 74–75
perceptual mapping and, 71–74
process illustration, 65
program, 64, 65
purpose, 64
routine and persistent, 63
trade shows and, 67
using, 68–71
valuable, 65
Competitive local exchange carriers, 3
Competitive markets model, 15–16
Competitive parity, 74–75
Competitive pricing, 119–20
Competitor proximity, 72
Concept testing, 46
Conjoint analysis, 166
Content management, 86
Continuous innovation, 171
Counterintelligence, 64
Credence goods, 184
Crises, 100
Crisis management, 100–102
challenge, 100
crisis plan, 101–2
example, 100–101
professionals, 101

reactive activities, 101
  See also Stakeholder management
Cross-channel strategies, 81–83
Customer-driven telecommunications marketing, 183–94
  need for, 1–2
  plan, 27–38
  uniqueness, 183–87
  See also Marketing
Customer lifetime value management, 113–15
  analysis, 114
  defined, 113
  example, 115
  goal, 113
  wireless life cycle example, 113–14
Customer profiling, 39–42
Customer relationship management (CRM), 42, 62, 143–55
  billing systems and, 147–50
  gamble, 144
  implementation, 145
  process support, 146
  support for churn management, 147
  systems, 143–47
  technology, 145
Customers
  account history, 140
  channel strategies for, 88
  decision-maker and, 8–11
  demand, 51–62
  directing marketing plan to, 29–32
  distribution channel and, 6–8
  lifetime value calculation, 140
  management process, 141
  on-line account management, 148
  regulatory environment and, 11
  as stakeholders, 96–97
  surveys, 45
  technology and, 12–13
  types of, 88
  view of competitors, 66
Customer self-service, 150, 150–53
  benefits, 150–51
  as cheapest delivery channel, 151
  coaxing, 153
  cost-effectiveness, 152
  customer dissatisfaction, 152
  elements, 150
  importance, 150
  savings, 151
  service provider challenges, 152

Database analytics, 136
Data mining, 39–42
  analytics, 40
  benefits, 40, 41
  defined, 40
  examples, 41–42
  as service provider assistance, 42
  software competition, 40
  technology, 42
Decision-makers
  as business buyers, 8, 9, 10
  consumer users as, 9
  customers and, 8–11
  structures, 10–11
Demand, 51–62
  business telecommunications, 59
  capital-intensive industry, 51–54
  curve, 56–58
  focus, 51–52
  technology industry, 55–56
Demand elasticity, 56–58
  limits, 57
  receptivity, 56
  unitary, 56
Demand forecasting, 58–62
  defined, 58
  ex post, 60
  improving, 61
  techniques, 59
  trend lines, 60
Demographic segmentation, 106
Design phase, 194
Deutsche Telekom, 4, 69
Differentiation, 20, 22–24
  attacking oneself and, 35–36
  attack strategies, 35–36
  creative pricing and, 23
  distribution channels and, 23, 79
  flanking strategies, 36
  as market coverage strategy, 111–12
  network quality and, 22
  opportunities, 22
Differentiation (continued)
  promotions and, 23
  strategies, 32–38
  value-added services and, 23

Diffusion of innovation curve, 12, 188–91
Direct network effects, 185
Discontinuous innovation, 171
Disintermediation, 78
Disruptive technologies
    adoption curve, 189
    development, 173–74
    as good marketers, 174
    identifying, 175
    losing market share to, 175–76
    marketing, 173–77
    successful, 177
    VoIP, 176
Distributed innovation, 171–73
Distribution channels
    agility, 79
    conflict, 6, 87–88
    cross-channel strategies, 81–83
    customers and, 6–8
    differentiation and, 23, 79
    lead costs, 83
    optimization, 81
    predestined, 7
    service providers and, 8
    streamlining, 89
    traditional, 77–78
DSL, 121
Dynamically continuous innovation, 171

Early adopters, 12–13
E-business channel management, 84–86
    challenges, 85
    on-line billing, 85
Economies of density, 53
Economies of scope, 53
Empathic design, 192–93
    customers and, 192
    for disruptive technology research, 193
    process, 193
Enterprise resource management (ERM) systems, 62
Environmental reports, 93
Exclusive club positioning, 163
Executive summary, 30
Ex post forecasting, 60
Focus groups, 45–46
Forecasting demand, 58–62
Framework Directive, 4
"Free-riding," 186
Free trials, 40

Gate theory pricing strategy, 128
Geographic segmentation, 105–6
Global Crossing, 37
Globalization, 93
Guerilla marketing, 37

Head-to-head positioning, 163

iHTML, 182
Indirect network effects, 185
Innovation
    bundling and, 25
    case study, 180–82
    centers, migration, 172
    continuous, 171
    customized, 172–73
    defined, 171
    diffusion curve, 188–91
    discontinuous, 171
    distributed, 171–73
    distribution and, 25
    dynamically continuous, 171
    marketing-based, 171–82
    organization, maintaining, 177–80
    pricing and, 25
    promotions and, 23
    support requirement, 177
    trialability, 178–79
Innovator's Dilemma, 173
Internet-based shopping/buying, 165
Issues management, 97–100
    defined, 97
    example, 99–100
    issue life cycle, 98
    privacy issue, 98
    proactive approach, 98
    process, 99
    stakeholder management vs., 99
    strategic issues, 99
    See also Stakeholder management

Late adopters, 13
Lead users, 191–92
Line extension, 168
Local number portability, 19, 185
Loyalty
    appreciation programs and, 137
    defined, 134–35
    maintenance strategies, 134–37
    retention vs., 134

reward programs and, 137
segmentation, 107
See also Churn

Marketing
   commodity, 164–65
   customer-driven, need for, 1–2
   defined, 24
   differentiating and, 22–24
   disruptive technologies, 173–77
   emerging opportunities, 24–25
   flanking, 36
   guerilla, 37
   one-to-one, 153–55
   particle, 109
   services as commodity, 20–22
   structure, 15–19
   technology services marketing, 183–87
   time displacement, 60
   tools, 191–94
Marketing-based innovation, 171–82
Marketing plan, 27–38
   advantages, 27–28
   customer-centered, 29, 32
   differentiation strategies, 32–38
   directing to customer, 29–32
   executive summary, 30
   length, 29–30
   objectives, 32
   planning process, 27–29
   previous, review, 28–29
   process, 29
   responsibility for completion, 28
   situation analysis, 30–31
   strategic approach, 31
   team spirit and, 28
Marketing Warfare, 33
Market research, 39–50
   concept testing, 46–47
   customer intercepts, 46
   customer profiling, 39–42
   customer surveys, 45
   data mining, 39–42
   focus groups, 45–46
   observational techniques, 46
   primary, 43–47
   secondary, 43–47
   yield management, 47–50
Markets
   concentration and, 112

   consumer, 113
   coverage strategies, 111–13
   penetration, 119
   undesirable, 116
   vertical, 112
Market segmentation, 105–16
   behavior, 106
   benefit, 108
   coverage statistics, 111–13
   customer lifetime value management, 113–14
   demographic, 106
   geographic, 105
   loyalty, 107
   markets/submarkets, 105–10
   particle marketing and, 109
   statistical, 110–11
   strategy evolution, 106
   undesirable markets and, 116
MCI, 21, 22, 122, 179
Message boards, 150
Monopoly services, 11

Natural language searches, 150
Network effects, 184–85
   direct, 185
   indirect, 185
Nextel, 123, 136
   market share growth, 34
   push-to-talk service, 74–75, 118, 166
NTT DoCoMo's i-Mode
   billing, 181
   case study, 180–82
   customer connection, 180
   "fun," 180
   iHTML, 182
   lock-in, 185
   menu-driven interface, 182
NTT DoCoMo's i-Mode (continued)
   open standards use, 181
   partnerships, 181

One-to-one marketing, 153–55
   concept, 153
   at logical limit, 154
   privacy concerns, 154–55
   See also Marketing
On-line billing, 85, 148
On-line communities, 153, 173

Packet-based pricing, 120–21
Particle marketing, 109, 153–55
Partner relationship management (PRM), 83–84
    defined, 83
    proliferation, 83
    security, 84
    system deployment, 84
    See also Channel management
Perceptual mapping, 71–74
    benefits, 72
    defined, 71
    example illustration, 73
Portals, 150
Preemptive positioning, 163
Price discrimination, 47–48, 125–27
    customers benefiting from, 127
    first-degree, 126
    occurrence, 125
    requirement, 125
    second-degree, 126
    third-degree, 126–27
Price elasticity, 56–58
Price leadership, 127–28
Pricing, 117–29
    broadband, 121–22
    bundling and, 123–25
    commodity, avoiding, 117–18
    competitive, 119–20
    for competitive parity, 127–29
    customer expectation and, 120
    differentiation and, 23
    discrimination, 125–27
    game theory strategy, 128
    innovation and, 25
    packet-based, 120–21
    promotional, 122–23
    skimming strategy, 119
    strategies, 119–22
Primary research, 43–46
    customer intercepts, 46
    customer surveys, 45
    focus groups, 45–46
    observational techniques, 46
    See also Market research
Product positioning, 162–64
    exclusive club, 163
    head-to-head, 163
    preemptive, 163

    repositioning, 163–64
    strategies, 162
    superlative, 163
    through product feature, 163
    See also Brand; Branding
Promotional pricing, 122–23
Promotions, 23

Quality function deployment (QFD), 193

Rebranding, 160
Regulatory environment, 11
Regulatory management, 97
Re-intermediation, 79
Retention, 134
Revenue management. *See* Yield management
Rewards programs, 137
Role-playing, 67

Secondary research, 43–47
Service providers
    branding, 118
    commodity, 21–22
    commodity buyer target, 21
    customer-centered, 7, 19
    customer self-service challenges, 152
    demand forecasting, 51
    distribution channels and, 8
    environmental reports, 93
    as low-cost provider, 20
    in multiple retail channels, 82
    in retail channel, 7
    in transforming market, 97
Service termination, 135–36
Signing bonuses, 135
Situation analysis, 30–31
    customer situation assessment, 30
    defined, 30
    service portfolio, 30
    technology assessment, 31
    See also Marketing plan
Skimming strategy, 119
Soft systems methodology (SSM), 95
Sprint, 67–68, 74
Stakeholder management, 91–102
    crisis management, 100–102
    defined, 93
    issues management, 97–100
    process illustration, 94

Stakeholders
  context, 92
  customers as, 96–97
  defined, 91
  importance of, 91–96
  partnerships with, 95
  power, 93–94
  scope of concerns, 95
  theory, 92
Statistical segmentation, 110–11
Strategic alliances, 86–87
  defined, 86
  examples, 86–87
  failure rate, 87
  life span, 87
  success factors, 87
  testing, 86
  See also Channel management
Strategic competitors, finding, 68–71
Strategic intelligence, 63–64
Strategic issues, 99
"Strategic square," 33
Superlative positioning, 163
Suppliers, power, 18

Tactical intelligence, 64
Technology
  customers and, 12–13
  disruptive, 173–77
  paradox, 185
  S-curve, 174
  sustaining, 174–75
Technology industry demand, 55–56
  management, 55–56
  pressure, 55
  See also Demand
Telecommunications
  as commodity, 117
  competition, 15–25

  equipment manufacturers, 192
  industry, 1–13
  marketing. *See* Marketing
  marketplace, state of, 3–6
Telecommunication services
  barrier to entry, 17
  differentiating, 20, 22–24
  marketing as commodity, 20–22
  monopoly, 11
Time displacement, 60
Tipping point, 13
T-Mobile, 34–35, 40, 80, 96
Trade shows, 67
Trialability, 178–79

Undesirable markets, 116
Usage pricing, 120–21

Value-added services, 23
Value proposition, 24
Verizon, 34, 74, 96, 123
VoIP, 71
  as disruptive technology, 176–77
  enhancements, 176
Vonage, 177, 180

Web-based self-provisioning, 79
Wi-Fi, hotspots growth, 4
Wireless application protocol (WAP), 80
Wireless market
  price leadership, 127
  U.S. market share, 34
WorldCom, 67–68, 102

Yield continuum, 49
Yield management, 47–50
  deep discounts and, 49
  defined, 47
  price discrimination, 47–48

Recent Titles in the Artech House Telecommunications Library

Vinton G. Cerf, Senior Series Editor

Access Networks: Technology and V5 Interfacing, Alex Gillespie

Achieving Global Information Networking, Eve L. Varma et al.

Advanced High-Frequency Radio Communications, Eric E. Johnson et al.

ATM Interworking in Broadband Wireless Applications, M. Sreetharan and S. Subramaniam

ATM Switches, Edwin R. Coover

ATM Switching Systems, Thomas M. Chen and Stephen S. Liu

Broadband Access Technology, Interfaces, and Management, Alex Gillespie

Broadband Local Loops for High-Speed Internet Access, Maurice Gagnaire

Broadband Networking: ATM, SDH, and SONET, Mike Sexton and Andy Reid

Broadband Telecommunications Technology, Second Edition, Byeong Lee, Minho Kang, and Jonghee Lee

The Business Case for Web-Based Training, Tammy Whalen and David Wright

Centrex or PBX: The Impact of IP, John R. Abrahams and Mauro Lollo

Chinese Telecommunications Policy, Xu Yan and Douglas Pitt

Communication and Computing for Distributed Multimedia Systems, Guojun Lu

Communications Technology Guide for Business, Richard Downey, Seán Boland, and Phillip Walsh

Community Networks: Lessons from Blacksburg, Virginia, Second Edition, Andrew M. Cohill and Andrea Kavanaugh, editors

Component-Based Network System Engineering, Mark Norris, Rob Davis, and Alan Pengelly

Computer Telephony Integration, Second Edition, Rob Walters

Customer-Centered Telecommunications Services Marketing, Karen G. Strouse

Deploying and Managing IP over WDM Networks, Joan Serrat and Alex Galis, editors

Desktop Encyclopedia of the Internet, Nathan J. Muller

Digital Clocks for Synchronization and Communications, Masami Kihara, Sadayasu Ono, and Pekka Eskelinen

Digital Modulation Techniques, Fuqin Xiong

E-Commerce Systems Architecture and Applications, Wasim E. Rajput

Engineering Internet QoS, Sanjay Jha and Mahbub Hassan

Error-Control Block Codes for Communications Engineers, L. H. Charles Lee

Essentials of Modern Telecommunications Systems, Nihal Kularatna and Dileeka Dias

FAX: Facsimile Technology and Systems, Third Edition, Kenneth R. McConnell, Dennis Bodson, and Stephen Urban

Fundamentals of Network Security, John E. Canavan

Gigabit Ethernet Technology and Applications, Mark Norris

Guide to ATM Systems and Technology, Mohammad A. Rahman

A Guide to the TCP/IP Protocol Suite, Floyd Wilder

Home Networking Technologies and Standards, Theodore B. Zahariadis

Information Superhighways Revisited: The Economics of Multimedia, Bruce Egan

Installation and Maintenance of SDH/SONET, ATM, xDSL, and Synchronization Networks, José M. Caballero et al.

Integrated Broadband Networks: TCP/IP, ATM, SDH/SONET, and WDM/Optics, Byeong Gi Lee and Woojune Kim

Internet E-mail: Protocols, Standards, and Implementation, Lawrence Hughes

Introduction to Telecommunications Network Engineering, Second Edition, Tarmo Anttalainen

Introduction to Telephones and Telephone Systems, Third Edition, A. Michael Noll

An Introduction to U.S. Telecommunications Law, Second Edition, Charles H. Kennedy

IP Convergence: The Next Revolution in Telecommunications, Nathan J. Muller

LANs to WANs: The Complete Management Guide, Nathan J. Muller

The Law and Regulation of Telecommunications Carriers,
  Henk Brands and Evan T. Leo

Managing Internet-Driven Change in International Telecommunications,
  Rob Frieden

*Marketing Telecommunications Services: New Approaches for a
  Changing Environment,* Karen G. Strouse

Mission-Critical Network Planning, Matthew Liotine

Multimedia Communications Networks: Technologies and Services,
  Mallikarjun Tatipamula and Bhumip Khashnabish, editors

Next Generation Intelligent Networks, Johan Zuidweg

Open Source Software Law, Rod Dixon

Performance Evaluation of Communication Networks,
  Gary N. Higginbottom

Performance of TCP/IP over ATM Networks, Mahbub Hassan and
  Mohammed Atiquzzaman

Practical Guide for Implementing Secure Intranets and Extranets,
  Kaustubh M. Phaltankar

Practical Internet Law for Business, Kurt M. Saunders

Practical Multiservice LANs: ATM and RF Broadband, Ernest O. Tunmann

Principles of Modern Communications Technology, A. Michael Noll

Programmable Networks for IP Service Deployment, Alex Galis et al.,
  editors

Protocol Management in Computer Networking, Philippe Byrnes

Pulse Code Modulation Systems Design, William N. Waggener

Security, Rights, and Liabilities in E-Commerce, Jeffrey H. Matsuura

Service Level Management for Enterprise Networks, Lundy Lewis

SIP: Understanding the Session Initiation Protocol, Second Edition,
  Alan B. Johnston

Smart Card Security and Applications, Second Edition, Mike Hendry

SNMP-Based ATM Network Management, Heng Pan

Spectrum Wars: The Policy and Technology Debate, Jennifer A. Manner

Strategic Management in Telecommunications, James K. Shaw

Strategies for Success in the New Telecommunications Marketplace,
  Karen G. Strouse

Successful Business Strategies Using Telecommunications Services,
  Martin F. Bartholomew

Telecommunications Cost Management, S. C. Strother

Telecommunications Department Management, Robert A. Gable

Telecommunications Deregulation and the Information Economy, Second Edition, James K. Shaw

Telecommunications Technology Handbook, Second Edition, Daniel Minoli

Telemetry Systems Engineering, Frank Carden, Russell Jedlicka, and Robert Henry

Telephone Switching Systems, Richard A. Thompson

Understanding Modern Telecommunications and the Information Superhighway, John G. Nellist and Elliott M. Gilbert

Understanding Networking Technology: Concepts, Terms, and Trends, Second Edition, Mark Norris

Videoconferencing and Videotelephony: Technology and Standards, Second Edition, Richard Schaphorst

Visual Telephony, Edward A. Daly and Kathleen J. Hansell

Wide-Area Data Network Performance Engineering, Robert G. Cole and Ravi Ramaswamy

Winning Telco Customers Using Marketing Databases, Rob Mattison

WLANs and WPANs towards 4G Wireless, Ramjee Prasad and Luis Muñoz

World-Class Telecommunications Service Development, Ellen P. Ward

For further information on these and other Artech House titles, including previously considered out-of-print books now available through our In-Print-Forever® (IPF®) program, contact:

Artech House
685 Canton Street
Norwood, MA 02062
Phone: 781-769-9750
Fax: 781-769-6334
e-mail: artech@artechhouse.com

Artech House
46 Gillingham Street
London SW1V 1AH UK
Phone: +44 (0)20 7596-8750
Fax: +44 (0)20 7630-0166
e-mail: artech-uk@artechhouse.com

Find us on the World Wide Web at:
www.artechhouse.com